T0250886

Plasma Physics and Controlled Thermonuclear Reactions Driven Fusion Energy

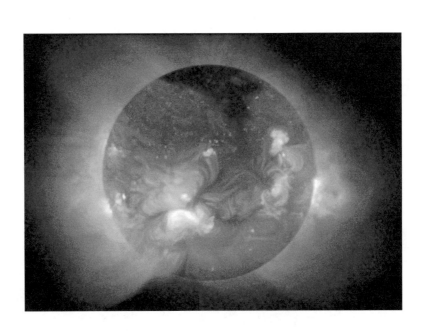

Bahman Zohuri

Plasma Physics and Controlled Thermonuclear Reactions Driven Fusion Energy

 Springer

Bahman Zohuri
Galaxy Advanced Engineering, Inc.
Albuquerque, New Mexico, USA

ISBN 978-3-319-83706-2   ISBN 978-3-319-47310-9   (eBook)
DOI 10.1007/978-3-319-47310-9

Printed on acid-free paper

This Springer imprint is published by Springer Nature
The registered company is Springer International Publishing AG
The registered company address is: Gewerbestrasse 11, 6330 Cham, Switzerland

This book is dedicated to my son Sasha.

Preface

Plasma is a unique state of matter, different from solids, liquids, and vapors. It is a gas where an important fraction of the atoms is ionized, so that the electrons and ions are separate, free, and consists of approximately equal numbers of positively charged ions and negatively charged electrons. The characteristics of plasmas are significantly different from those of ordinary neutral gases so that plasmas are considered a distinct "fourth state of matter." For example, because plasmas are made up of electrically charged particles, they are strongly influenced by electric and magnetic fields, while neutral gases are not. An example of such influence is the trapping of energetic charged particles along geomagnetic field lines to form the Van Allen radiation belts.

Plasma physics is the study of charged particles and fluids interacting with self-consistent electric and magnetic fields. It is a basic research discipline that has many different areas of application—space and astrophysics, controlled fusion, accelerator physics, and beam storage, to name a few.

In addition to externally imposed fields, such as the Earth's magnetic field or the interplanetary magnetic field, plasma is acted upon by electric and magnetic fields created within the plasma itself through localized charge concentrations and electric currents that result from the differential motion of the ions and electrons. The forces exerted by these fields on the charged particles that make up the plasma act over long distances and impart to the particles' behavior a coherent, collective quality that neutral gases do not display. Despite the existence of localized charge concentrations and electric potentials, plasma is electrically "quasi-neutral," because, in aggregate, there are approximately equal numbers of positively and negatively charged particles distributed so that their charges cancel.

Plasma science has, in turn, spawned new avenues of basic science. Most notably, plasma physicists were among the first to open up and develop the new and profound science of chaos and nonlinear dynamics. Plasma physicists have also contributed greatly to studies of turbulence, which is important for safe air travel. Basic plasma science continues to be a vibrant research area. Recent new discoveries have occurred in understanding extremely cold plasmas, which condense to crystalline states, the study of high-intensity laser interactions, new highly efficient lighting systems, and plasma-surface interactions important for computer manufacturing.

Understanding the complex behavior of confined plasmas has led researchers to formulate the fundamental equations of plasma physics. This foundational work and understanding of plasmas have led to important advances in fields as diverse as computers, lighting, waste handling, space physics, switches and relays, and lasers. Because plasmas are conductive, respond to electric and magnetic fields, and can be efficient sources of radiation, they are used in a large number of applications where such control is needed or when special sources of energy or radiation are required.

The book is intended only as an introduction to plasma physics course and includes what I take to be the critical concepts needed for a foundation for further study. A solid undergraduate background in classical physics and electromagnetic theory including Maxwell's equations and mathematical familiarity with partial differential equations and complex analysis are prerequisites.

In summary, it is clear that this book is not intended to transform its users into experts on plasma physics. Rather, it is intended to provide a simple, coherent, introduction to workers with diverse backgrounds in physics and related sciences.

Albuquerque, NM   Bahman Zohuri

Acknowledgements

I am indebted to the many people who aided, encouraged, and supported me. Some are not around to see the results of their encouragement in the production of this book, yet I hope they know of my deepest appreciations. I especially want to thank my friend Masoud Moghadam, to whom I am deeply indebted, who has continuously given his support without hesitation. He has always kept me going in the right direction. His intelligent comments and suggestion to different chapters and sections made this book available. Above all, I offer very special thanks to my late mother and father and to my children, in particular my son Sasha. They have provided constant interest and encouragement, without which this book would not have been written. Their patience with my many absences from home and long hours in front of the computer to prepare the manuscript is especially appreciated.

Contents

About the Author

Dr. Bahman Zohuri currently works for Galaxy Advanced Engineering, Inc., a consulting firm that he started in 1991 when he left both the semiconductor and defense industries after many years working as a chief scientist. After graduating from the University of Illinois in the field of physics and applied mathematics, he then went to the University of New Mexico, where he studied nuclear engineering and mechanical engineering. He joined Westinghouse Electric Corporation, where he performed thermal hydraulic analysis and studied natural circulation in an inherent shutdown heat removal system (ISHRS) in the core of a liquid metal fast breeder reactor (LMFBR) as a secondary fully inherent shutdown system for secondary loop heat exchange. All these designs were used in nuclear safety and reliability engineering for a self-actuated shutdown system. He designed a mercury heat pipe and electromagnetic pumps for large pool concepts of a LMFBR for heat rejection purposes for this reactor around 1978, when he received a patent for it. He was subsequently transferred to the defense division of Westinghouse, where he oversaw dynamic analysis and methods of launching and controlling MX missiles from canisters. The results were applied to MX launch seal performance and muzzle blast phenomena analysis (i.e., missile vibration and hydrodynamic shock formation). Dr. Zohuri was also involved in analytical calculations and computations in the study of nonlinear ion waves in rarefying plasma. The results were applied to the propagation of so-called soliton waves and the resulting charge collector traces in the rarefaction characterization of the corona of laser-irradiated target pellets. As part of his graduate research work at Argonne National Laboratory, he performed computations and programming of multi-exchange integrals in surface physics and solid-state physics. He earned various patents in areas such as diffusion processes and diffusion furnace design while working as a senior process engineer at various semiconductor companies, such as Intel Corp., Varian Medical Systems, and National Semiconductor Corporation. He later joined Lockheed Martin Missile and Aerospace Corporation as senior chief scientist and oversaw research and development (R&D) and the study of the vulnerability, survivability, and both radiation and laser hardening of different components of the Strategic Defense Initiative, known as Star Wars.

This included payloads (i.e., IR sensor) for the Defense Support Program, the Boost Surveillance and Tracking System, and Space Surveillance and Tracking Satellite against laser and nuclear threats. While at Lockheed Martin, he also performed analyses of laser beam characteristics and nuclear radiation interactions with materials, transient radiation effects in electronics, electromagnetic pulses, system-generated electromagnetic pulses, single-event upset, blast, thermomechanical, hardness assurance, maintenance, and device technology.

He spent several years as a consultant at Galaxy Advanced Engineering serving Sandia National Laboratories, where he supported the development of operational hazard assessments for the Air Force Safety Center in collaboration with other researchers and third parties. Ultimately, the results were included in Air Force Instructions issued specifically for directed energy weapons operational safety. He completed the first version of a comprehensive library of detailed laser tools for airborne lasers, advanced tactical lasers, tactical high-energy lasers, and mobile/tactical high-energy lasers, for example.

He also oversaw SDI computer programs, in connection with Battle Management CI and artificial intelligence and autonomous systems. He is the author of several publications and holds several patents, such as for a laser-activated radioactive decay and results of a through-bulkhead initiator. He has published the following works: *Heat Pipe Design and Technology: A Practical Approach* (CRC Press), *Dimensional Analysis and Self-Similarity Methods for Engineers and Scientists* (Springer), *High Energy Laser (HEL): Tomorrow's Weapon in Directed Energy Weapons Volume I* (Trafford Publishing Company), and recently the book on the subject directed energy weapons and physics of high-energy laser with Springer. He has other books with Springer Publishing Company: *Thermodynamics in Nuclear Power Plant Systems* (Springer) and *Thermal-Hydraulic Analysis of Nuclear Reactors* (Springer).

Chapter 1
Foundation of Electromagnetic Theory

In order to study plasma physics and its behavior for a source of driving fusion for a controlled thermonuclear reaction for the purpose of generating energy, understanding of the fundamental knowledge of electromagnetic theory is essential. In this chapter, we introduce Maxwell's equations and the Coulomb barrier or tunnel effects for better understanding of plasma behavior for confinement purpose of controlled thermonuclear reaction and for generating clean energy that is confined magnetically in particular. We are mainly concern with confinement of plasmas at terrestrial temperature, e.g., very hot plasmas, where primarily of interest is in the application to controlled fusion research in magnetic confinement reactors such as tokomak.

1.1 Introduction

Although Maxwell's equation was formulated by him over 100 decades ago, the subject of electromagnetism was never stagnated. The production of the so-called clean energy is driven by magnetic confinement of hot plasma via controlled thermonuclear reaction between two isotopes of hydrogen, namely, deuterium (D) and tritium (T), resulting in some behavior in plasma that is known as magnetohydrodynamics abbreviated as MHD. The study of such phenomena requires knowledge and understanding of the fundamental of electromagnetisms and fluid dynamics combined where *fluid dynamics equation and Maxwell's equation are coupled.*

However, in the study of electricity and magnetism, as part of understanding the physics of plasma, we need to have some knowledge of notation that may be accomplished by using the notation of vector analysis. To provide the valuable and shorthanded electromagnetic and electrodynamics, vector analysis also brings to the forefront the physical ideas involved in these equations; therefore, we will

© Springer International Publishing AG 2016
B. Zohuri, *Plasma Physics and Controlled Thermonuclear Reactions Driven Fusion Energy*, DOI 10.1007/978-3-319-47310-9_1

briefly formulate some of these vector analysis concepts and present some of their identity in this chapter.

1.2 Vector Analysis

In the study of fundamental science of physics, several kinds of quantities are encountered; in particular, we need to distinguish *vectors* and *scalars*. For our purposes, it is sufficient to define a scalar as follows:

1. **Scalar**: A *scalar* is a quantity that is completely characterized by its magnitude. Examples of scalars are mass, volume, etc. A simple extension of the idea of a scalar is a *scalar field*—a function of position that is completely specified by its magnitude at all points in space.
2. **Vector**: A *vector* is a quantity that is completely characterized by its magnitude and direction. Examples of vectors are that we consider position from a fixed origin, velocity, acceleration, force, etc. The generalization to a *vector field* gives a function of position that is completely specified by its magnitude and direction at all points in space.

The detailed analysis of vector analysis is beyond the scope of this book; thus, we will briefly formulate the fundamental layout of vector analysis here for the purpose of vector analysis operation and operator developing essential electromagnetic and electrodynamics that are the foundation for understanding of plasma physics.

1.2.1 Vector Algebra

We are familiar with scalar algebra from our basic algebra courses and some algebra can be applied to develop vector algebra as well. For the time being, we use the Cartesian coordinate system to develop the three-dimensional analysis of vector algebra. The Cartesian system allows to represent a vector by its three components and denote them by x, y, and z, or when it is more convenient, we use notation of x_1, x_2, and x_3. With respect to the Cartesian coordinate system, a vector is specified by its x-, y-, and z-components. Thus, a vector $\vec{V}$ (note that the vector quantities are denoted by symbol of vector $\rightarrow$ on top) is specified by its components, V_x, V_y, and V_z, where $V_x = |\vec{V}| \cos \alpha_1$, $V_y = |\vec{V}| \cos \alpha_2$, and $V_z = |\vec{V}| \cos \alpha_3$, the αs being the angles between vector $\vec{V}$ and the appropriate coordinate axes of the Cartesian system. The scalar $|\vec{V}| = \sqrt{V_x^2 + V_y^2 + V_z^2}$ is the *magnitude* of the vector or its length. Utilizing Fig. 1.1, in the case of vector fields, each of the components is to be regarded as a function of x, y, and z. It should be

Fig. 1.1 Presentation of
vector along with its
components in the Cartesian
coordinate system

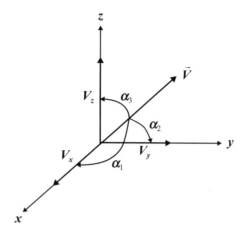

emphasized that for the simplicity of analysis, we are using the Cartesian coordinate system, yet the similarity of these analyses applies to the other coordinates such as cylindrical and spherical coordinates, respectively, as well.

1.2.1.1   Sum of Two Vectors

The sum of two vectors $\vec{A}$ and $\vec{B}$ is defined as the vector $\vec{C}$ whose components are the sum of corresponding components of the original vectors. Thus, we can write:

$$\vec{C} = \vec{A} + \vec{B} \tag{1.1}$$

and

$$\begin{aligned}
C_x &= A_x + B_x \\
C_y &= A_y + B_y \\
C_z &= A_z + B_z
\end{aligned} \tag{1.2}$$

This definition of the vector sum is completely equivalent to the familiar parallelogram rule for vector addition.

1.2.1.2   Subtraction of Two Vectors

Vector subtraction is defined in terms of the negative of a vector, which is the vector whose components are the negative of the corresponding components of the original vector. Thus, if $\vec{A}$ is a vector, $-\vec{A}$ is defined by:

$$\begin{aligned}
(-A)_x &= -A_x \\
(-A)_y &= -A_y \\
(-A)_z &= -A_z
\end{aligned} \tag{1.3}$$

The operation of subtraction is then defined as the addition of the negative and is written as:

$$\vec{A} - \vec{B} = \vec{A} + \left(-\vec{B}\right) \tag{1.4}$$

Since the addition of real numbers is associative and commutative, it follows that vector addition and subtraction are also associative and commutative. In vector form notation, this appears as:

$$\begin{aligned}
\vec{A} + \left(\vec{B} + \vec{C}\right) &= \left(\vec{A} + \vec{B}\right) + \vec{C} \\
&= \left(\vec{A} + \vec{C}\right) + \vec{B} \\
&= \vec{A} + \vec{B} + \vec{C}
\end{aligned} \tag{1.5}$$

In other words, the parentheses are not needed as indicated by the last form.

1.2.1.3 Multiplication of Two Vectors

Now, we proceed to multiplication of two vectors and their process. We note that the simplest product is a scalar times a vector. This operation results in a vector, each component of which is the scalar times the corresponding component of the original vector. If c is a scalar and $\vec{A}$ is a vector and the product $c\vec{A}$ is a vector, then $\vec{B} = c\vec{A}$ is defined by:

$$\begin{aligned}
B_x &= cA_x \\
B_y &= cA_y \\
B_z &= cA_z
\end{aligned} \tag{1.6}$$

It is clear that if $\vec{A}$ is a *vector field* and c is a *scalar field*, then $\vec{B}$ is a new vector field that is *not* necessary a constant multiple of the origin field.

If we like to multiply two vectors together, there are two possibilities and they are known as the *vector* and *scalar* product or sometimes they are called *cross* or *dot* products, respectively.

Scalar Product of Two Vectors

First considering the scalar or dot product of two vectors $\vec{A}$ and $\vec{B}$, we note that sometimes the scalar product called *inner product* is derived from the scalar nature of the product. The definition of the scalar product is written as:

$$\vec{A} \cdot \vec{B} = A_x B_x + A_y B_y + A_z B_z \tag{1.7}$$

This definition is equivalent to another, and perhaps more familiar, definition—that is, as the product of the magnitudes of the original vectors times the cosine of the angle between these vectors. If they are perpendicular to each other:

$$\vec{A} \cdot \vec{B} = 0 \tag{1.8}$$

Note that the scalar product is commutative. The length of $\vec{A}$ is then:

$$|\vec{A}| = \sqrt{\vec{A} \cdot \vec{A}} \tag{1.9}$$

Vector Product of Two Vectors

The vector product of two vectors is a vector, which accounts for the name and alternative names *outer product* and *cross product*. The vector product is written as $\vec{A} \times \vec{B}$. If $\vec{C}$ is the vector product of $\vec{A}$ and $\vec{B}$, then:

$$\vec{C} = \vec{A} \times \vec{B} \tag{1.10}$$

or in terms of their components can be written as:

$$\begin{aligned} C_x &= A_y B_z - A_z B_y \\ C_y &= A_z B_x - A_x B_z \\ C_z &= A_x B_y - A_y B_x \end{aligned} \tag{1.11}$$

It is important to note that the cross product depends on the order of the factors; interchanging the order of the cross product introduces a minus sign as:

$$\vec{B} \times \vec{A} = -\vec{A} \times \vec{B} \tag{1.12}$$

Consequently:

$$\vec{A} \times \vec{A} = 0 \tag{1.13}$$

Fig. 1.2 Right-hand
screw rule

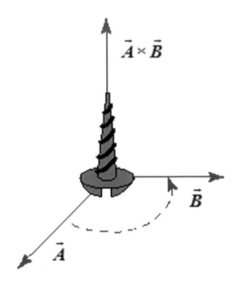

This definition is equivalent to the following: the vector product is the product of the magnitudes times the sine of the angle between the original vectors, with the direction given by the right-hand screw rule (see Fig. 1.2 here).

Note that if we let $\vec{A}$ be rotated into $\vec{B}$ through the smallest possible angle, a right-hand screw rotated in this manner will advance in a direction perpendicular to both $\vec{A}$ and $\vec{B}$; this direction is the direction of $\vec{A} \times \vec{B}$.

The vector product may be easily expressed in terms of a determinant via definition of unit vectors as $\hat{\imath}, \hat{\jmath}$, and $\hat{k}$, which are vectors of unit magnitude in the x-, y-, and z-directions, respectively; then we can write:

$$\vec{A} \times \vec{B} = \begin{vmatrix} \hat{\imath} & \hat{\jmath} & \hat{k} \\ A_x & A_y & A_z \\ B_x & B_y & B_z \end{vmatrix} \tag{1.14}$$

If this determinant is evaluated by the usual rules, the result is precisely our definition of the cross product of two vectors.

The determinant in Equation 1.14 may be combined in many ways and most of the results that are obtained are obvious; however, there are two triple products of sufficient importance that need to be mentioned. The triple scar product $D = \vec{A} \cdot \vec{B} \times \vec{C}$ is easily found to be given by the determinant as:

$$D = \vec{A} \cdot \vec{B} \times \vec{C} = \begin{vmatrix} A_x & A_y & A_z \\ B_x & B_y & B_z \\ C_x & C_y & C_z \end{vmatrix} = -\vec{B} \cdot \vec{A} \times \vec{C} \tag{1.15}$$

This product in Equation 1.15 is unchanged by an exchange of dot and cross or by a cyclic permutation of the three vectors. Note that parentheses are not needed, since the cross product of a scalar and a vector is undefined.

The other interesting triple product is the triple vector product $\vec{D} = \vec{A} \times \left(\vec{B} \times \vec{C} \right)$. By a repeated application of the definition of the cross product, Equations 1.10 and 1.11, we find that:

$$\vec{D} = \vec{A} \times \left(\vec{B} \times \vec{C} \right) = \vec{B} \left(\vec{A} \cdot \vec{C} \right) - \vec{C} \left(\vec{A} \cdot \vec{B} \right) \qquad (1.16)$$

which is frequently known as the *back cab rule*. We should bear in mind that in the cross product, the parentheses are vital as part of the operation and without them the product is not well defined.

1.2.1.4 Division of Two Vectors

At this point one might be interested as to the possibility of vector division. Division of a vector by scalar can, of course, be defined as multiplication by the reciprocal of the scalar. Division of a vector by another vector, however, is possible only if the two vectors are parallel. On the other hand, it is possible to write general solution to vector equations and so accomplish so meting closely akin to division. Consider the equation below as:

$$c = \vec{A} \cdot \vec{X} \qquad (1.17)$$

where c is a known scalar. $\vec{A}$ is a known vector, and $\vec{X}$ is an unknown vector. A general solution to the Equation 1.17 is given as follows:

$$\vec{X} = \frac{c\vec{A}}{\vec{A} \cdot \vec{A}} + \vec{B} \qquad (1.18)$$

where $\vec{B}$ is an arbitrary vector that is perpendicular to $\vec{A}$ and that is $\vec{A} \cdot \vec{B} = 0$. What we have done is very nearly to divide c by vector $\vec{A}$; more correctly, we have found the general form of the vector $\vec{X}$ that satisfies Equation 1.17. There is no unique solution, and this fact accounts for the vector $\vec{B}$. In the same fashion, we may consider the vector equation as:

$$\vec{C} = \vec{A} \times \vec{X} \qquad (1.19)$$

In Equation 1.19, both vectors $\vec{A}$ and $\vec{C}$ are known vectors and $\vec{X}$ is an unknown vector. The general solution of this equation is then given by:

$$\vec{X} = \frac{\vec{C} \times \vec{A}}{\vec{A} \cdot \vec{A}} + k\vec{A} \qquad (1.20)$$

where k is an arbitrary scalar. Thus, $\vec{X}$ as defined by Equation 1.20 is very nearly the quotient of $\vec{C}$ by $\vec{A}$; the scalar k takes account of the nonuniqueness of the process. If $\vec{X}$ is required to satisfy both Equations 1.17 and 1.19, then the result is unique, if it exists and is given by:

$$\vec{X} = \frac{\vec{C} \times \vec{A}}{\vec{A} \cdot \vec{A}} + \frac{c\vec{A}}{\vec{A} \cdot \vec{A}} \qquad (1.21)$$

1.2.2 Vector Gradient

Now that we have covered basic vector algebra, we pay our attention to vector calculus, which extends to vector gradient, integration, vector curl, and differentiation of vectors. The simplest of these is the relation of a particular vector field to the derivative of a scalar field.

For that matter, it is convenient to introduce the idea of *directional derivative* of a function of several variables, which we leave it to the reader to find these analyses in any vector calculus book to find the details of such derivative that is beyond the intended scope of this book, and we just jump to the definition of vector gradient.

The gradient of a scalar function φ is a vector whose magnitude is the maximum directional derivative at the point is being considered and whose direction is the direction of the maximum directional derivative at the point. Using the geometry of Fig. 1.3, we put this definition into some perspective, and it is evident that the gradient has the direction to the level surface of φ through the point as we said is being coinsured.

Fig. 1.3 Parts of two level surfaces of the function $\varphi(x, y, z)$

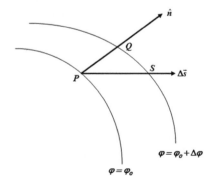

The most common mathematical symbol for gradient is $\vec{\nabla}$ or in text form is **grad**. In terms of the gradient, the directional derivative is given by:

$$\frac{d\varphi}{ds} = |\mathrm{grd}\,\vec{\varphi}| \cos \theta \tag{1.22}$$

where θ is the angle between the direction of $d\vec{s}$ and the direction of the gradient. This result is immediately evident from Fig. 1.3. If we write $d\vec{s}$ for the vector displacement of magnitude ds, then Equation 1.22 can be written as:

$$\frac{d\varphi}{ds} = \mathrm{grd}\,\vec{\varphi} . \frac{d\vec{s}}{ds} \tag{1.23}$$

Equation 1.23 enables us to seek for the explicit form of the gradient and find that in any coordinate system in which we know the form of $d\vec{s}$. In the Cartesian or rectangular coordinate system, we know that $d\vec{s} = \hat{i}\,dx + \hat{j}\,dy + \hat{k}\,dz$. We also know from differential calculus that:

$$d\varphi = \frac{\partial \varphi}{\partial x}dx + \frac{\partial \varphi}{\partial y}dy + \frac{\partial \varphi}{\partial z}dz \tag{1.24}$$

From Equation 1.22, it results that:

$$\begin{aligned} d\varphi \ &= \frac{\partial \varphi}{\partial x}dx + \frac{\partial \varphi}{\partial y}dy + \frac{\partial \varphi}{\partial z}dz \\ &= (\mathrm{grd}\varphi)_x dx + (\mathrm{grd}\varphi)_y dy + (\mathrm{grd}\varphi)_z dz \end{aligned} \tag{1.25}$$

Equating coefficient of independent variables on both sides of the equation in rectangular coordinate, it gives:

$$\mathrm{grd}\,\vec{\varphi} = \hat{i}\,\frac{\partial \varphi}{\partial x} + \hat{j}\,\frac{\partial \varphi}{\partial y} + \hat{k}\,\frac{\partial \varphi}{\partial z} \tag{1.26}$$

In a more complicated case, the procedure is very similar as well. In spherical polar coordinates with utilization of Fig. 1.4 with denotation of r, θ, and ϕ, we can write Equation 1.24 in the following form as:

$$d\varphi = \frac{\partial \varphi}{\partial r}dr + \frac{\partial \varphi}{\partial \theta}d\theta + \frac{\partial \varphi}{\partial \phi}d\phi \tag{1.27}$$

and

$$d\vec{s} = \hat{a}_r dr + \hat{a}_\theta r d\theta + \hat{a}_\phi r \sin \theta d\phi \tag{1.28}$$

Fig. 1.4 Definition of the
polar coordinates

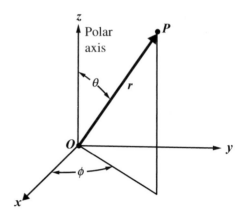

where $\hat{a}_r$, $\hat{a}_\theta$, and $\hat{a}_\phi$ are unit vectors in the r, θ, and ϕ directions, respectively. Applying Equation 1.23 and equating coefficients of independent variables yield:

$$\mathrm{grd}\,\vec{\varphi} = \hat{a}_r \frac{\partial \varphi}{\partial r} + \hat{a}_\theta \frac{1}{r} \frac{\partial \varphi}{\partial \theta} + \hat{a}_\phi \frac{1}{r \sin \theta} \frac{\partial \varphi}{\partial z} \tag{1.29}$$

Equation 1.29 is established in spherical coordinate system.

1.2.3 Vector Integration

Although there are other aspects of vector differentiation first, we need to discuss the vector integration, and details of such analyses are left to the reader to look them up in any vector calculus book and just briefly formulate them here. For purposes of vector integration, we will consider three kinds of integrals, according to the nature of the differential appearing in integral, and they are:

1. Line integral
2. Surface integral
3. Volume integral

In either case, the integrand may be either a vector or a scalar field; however, certain combinations of integrands and differentials give rise to uninteresting integrals. Those of most interest here are the scalar line integral of a vector, the scalar surface integral of a vector, and finally the volume integral of both vectors and scalars.

If $\vec{F}$ is a vector field, a line integral of $\vec{F}$ is written as:

$$\int_{a(C)}^{b} \vec{F}(\vec{r}) \cdot \mathrm{d}\vec{l} \tag{1.30}$$

where C is the curve along which the integration is performed, a and b are the initial and final points on the curve, and $d\vec{l}$ is an infinitesimal vector displacement along the curve C.

It is obvious since the result of dot product of $\vec{F}(\vec{r}) \cdot d\vec{l}$ is scalar, then the result of linear integral in Equation 1.30 is scalar. The definition of line integral follows closely the Riemann definition of the definite integral; thus, the integral can be written as segment of curve C between the lower and upper bound of a and b, respectively, that can be divided into a large number of small increments $\Delta\vec{l}$; for increment an interior point is chosen and the value of $\vec{F}(\vec{r})$ at that point is found.

In other words, Equation 1.30 can form the following form of equation as:

$$\int_{a(C)}^{b} \vec{F}(\vec{r}) \cdot d\vec{l} = \lim_{N \to \infty} \sum_{i=1}^{N} \vec{F}_i(\vec{r}) \cdot \Delta\vec{l} \qquad (1.31)$$

It is important to emphasize that the line integral usually depends not only on the endpoint a and b but also on the curve C along which the integration is to be done, since the magnitude and direction of $\vec{F}(\vec{r})$ and the direction of $d\vec{l}$ depend on curve C and its tangent, respectively. The line integral around a closed curve is of sufficient importance that a special notation is used for it, namely:

$$\oint_{C} \vec{F} \cdot d\vec{l} \qquad (1.32)$$

Note that the integral around a closed curve is usually not zero. The class of vectors for which the line integral around any closed curve is zero is of considerable importance. Thus, we normally write line integrals around undesignated closed paths as:

$$\oint \vec{F} \cdot d\vec{l} \qquad (1.33)$$

The form of integral in Equation 1.33 around closed curve C is for those cases where the integral is independent of the contour C within rather wide limits.

Now paying our attention to the second kind of integral, namely, surface integral, we can again define $\vec{F}$ as a vector; a surface integral of $\vec{F}$ is written as:

$$\int_{S} \vec{F} \cdot \hat{n} \, da \qquad (1.34)$$

where S is the surface over which the integral is taken, da is an infinitesimal area on surface S, and $\hat{n}$ is the unit vector normal to da.

There are two degrees of ambiguity in the choice of unit vector $\hat{n}$ as far as outward or downward direction normal to surface S is concerned, if this surface is a

Fig. 1.5 Relation of normal
unit vector to the surface
and the direction of
traversal of the boundary

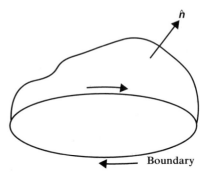

closed one. If S is not closed and is finite, then it has a boundary, and the sense of the normal is important only with respect to the arbitrary positive sense of traversing the boundary. The positive sense of the normal is the direction in which a right-hand screw would advance if rotated in the direction of the positive sense on the bounding curve, as it is illustrated in Fig. 1.5. The surface integral of $\vec{F}$ over a closed surface S is sometimes denoted by:

$$\oint_S \vec{F} \cdot \hat{n}\,\mathrm{d}a \tag{1.35}$$

Comments exactly parallel to those made for the line integral can be made for the surface integral. This surface integral is clearly a scalar and it usually depends on the surface S, and cases where it does not are particularly important.

Now, we can pay our attention to the third type of vector integral, namely, volume integral, and again we start with vector $\vec{F}$. Therefore, if $\vec{F}$ is a vector and φ is a scalar, then the two volume integrals in which we are interested are written as:

$$J = \int_V \varphi\,\mathrm{d}\upsilon \qquad \vec{K} = \int_V \vec{F}\,\mathrm{d}\upsilon \tag{1.36}$$

Clearly J is a scalar and $\vec{K}$ is a vector. The definitions of these integrals reduce quickly to just the Riemann integral in three dimension except that in $\vec{K}$ one must note that there is one integral for each component of $\vec{F}$. However, we are very familiar with these integrals and require no further investigation nor any comments.

1.2.4 Vector Divergence

Another important vector operator, which is playing an essential role in establishing electromagnetism equations, is vector divergence operation, which is a derivative form. The divergence of vector $\vec{F}$, written as div $\vec{F}$, is defined as follows.

The divergence of a vector is the limit of its surface integral per unit volume as the volume enclosed by the surface goes to zero. This statement mathematically can be presented as follows:

$$\text{div}\,\vec{F} = \lim_{V \to 0} \frac{1}{V} \oint_S F \cdot \hat{n}\, da \tag{1.37}$$

The divergence is clearly a scalar point function and its result of operation ends up with scalar field, and it is defined at the limit point of the surface of integration. A detail of proof of this concept is beyond the scope of the book and it is left to the reader to refer to any vector calculus book. However, the limit is easily can be taken, and the divergence in rectangular coordinates is found to be:

$$\text{div}\,\vec{F} = \frac{\partial F_x}{\partial x} + \frac{\partial F_y}{\partial y} + \frac{\partial F_z}{\partial z} \tag{1.38}$$

Equation 1.38 for vector divergence operation designated for the Cartesian coordinate and in spherical coordinate is written in the following form:

$$\text{div}\,\vec{F} = \frac{1}{r^2} \frac{\partial}{\partial r}\left(r^2 F_r\right) + \frac{1}{r \sin\theta} \frac{\partial}{\partial \theta}\left(\sin\theta F_\theta\right) + \frac{1}{r \sin\theta} \frac{\partial F_\phi}{\partial \phi} \tag{1.39}$$

and in cylindrical coordinate is presented by:

$$\text{div}\,\vec{F} = \frac{1}{r} \frac{\partial}{\partial r}\left(r F_r\right) + \frac{1}{r} \frac{\partial}{\partial \theta}\left(F_\theta\right) + \frac{\partial}{\partial z}\left(F_z\right) \tag{1.40}$$

The method of finding the explicit of the divergence is applicable to any coordinate system, provided that the forms of the volume and surface elements or, alternatively, the elements of the length are known.

Now that we have the idea behind the vector divergence operator and its operation, we can then establish the *divergence theorem*. The integral of the divergence of a vector over a volume V is equal to the surface integral of the normal component of the vector over the surface bounding V, that is:

$$\int_V \text{div}\,\vec{F}\, dv = \oint_S \vec{F} \cdot \hat{n}\, da \tag{1.41}$$

and we leave it as that, and again for proof one can refer to any vector calculus book.

1.2.5   Vector Curl

Another interesting vector differential operator is the vector curl. The curl of a vector, written as curl $\vec{F}$, is defined as follows.

The *curl* of a vector is the limit of the ratio of the integral of its cross product with the outward drawn normal, over a closed surface, to the volume enclosed by the surface as the volume goes to zero, that is:

$$\text{curl } \vec{F} = \lim_{V \to 0} \frac{1}{V} \oint_S \hat{n} \times \vec{F} da \qquad (1.42)$$

Again the details of proof are left to the reader to find them out in a vector calculus book, and we just write the final result of curl operator as follows in at least rectangular coordinate:

$$\text{curl } \vec{F} = \begin{vmatrix} \hat{i} & \hat{j} & \hat{k} \\ \dfrac{\partial}{\partial x} & \dfrac{\partial}{\partial y} & \dfrac{\partial}{\partial z} \\ F_x & F_y & F_z \end{vmatrix} \qquad (1.43)$$

Finding the form of the curl in other coordinate system is only slightly more complicated and it is left to the reader for practice.

Now that we have understanding of the vector curl operator, we can now state *Stock's theorem* as follows.

The line integral of a vector around a closed curve is equal to the integral of the normal component of its curl over any surface bounded by the curve, that is:

$$\oint_C \vec{F} \cdot d\vec{l} = \int_S \text{curl } \vec{F} \cdot \hat{n} \, da \qquad (1.44)$$

where C is a closed curve that bounds the surface S.

1.2.6   Vector Differential Operator

We now introduce an alternative notation for the types of vector differentiation that have been discussed—namely, gradient, divergence, and curl. This notation uses the vector differential operator *del* and it is identified as symbol of $\vec{\nabla}$ and mathematically written as:

$$\vec{\nabla} = \hat{i}\,\frac{\partial}{\partial x} + \hat{j}\,\frac{\partial}{\partial y} + \hat{k}\,\frac{\partial}{\partial z} \tag{1.45}$$

Del is a differential operator in that it is used only in front of a function of (x, y, z), which it differentiates; it is a vector in that it obeys the laws of vector algebra. It is also a vector in terms of its transformation properties and in terms of del Equations 1.46, 1.47, and 1.48 are expressed as follows:

Grad $= \vec{\nabla}$:

$$\vec{\nabla}\varphi = \hat{i}\,\frac{\partial \varphi}{\partial x} + \hat{j}\,\frac{\partial \varphi}{\partial y} + \hat{k}\,\frac{\partial \varphi}{\partial z} \tag{1.46}$$

Div $= \vec{\nabla}\cdot$:

$$\vec{\nabla} \cdot \vec{F} = \frac{\partial F_x}{\partial x} + \frac{\partial F_y}{\partial y} + \frac{\partial F_z}{\partial z} \tag{1.47}$$

Curl $= \vec{\nabla}\times$:

$$\vec{\nabla} \times \vec{F} = \begin{vmatrix} \hat{i} & \hat{j} & \hat{k} \\ \dfrac{\partial}{\partial x} & \dfrac{\partial}{\partial y} & \dfrac{\partial}{\partial z} \\ F_x & F_y & F_z \end{vmatrix} \tag{1.48}$$

The operations expressed with del are themselves independent of any special choice of coordinate system. Moreover, any identities that can be proved using the Cartesian representation hold independently of the coordinate system.

1.3   Further Developments

The first of these is the *Laplacian operator,* which is defined as the divergence of the gradient of a scalar field and which is usually written as ∇^2:

$$\vec{\nabla} \cdot \vec{\nabla} = \nabla^2 \tag{1.49}$$

In rectangular coordinates:

$$\nabla^2 \varphi = \frac{\partial^2 \varphi}{\partial x^2} + \frac{\partial^2 \varphi}{\partial y^2} + \frac{\partial^2 \varphi}{\partial z^2} \tag{1.50}$$

This operator is of great importance in electrostatics and will be considered in the following sections and chapters.

The curl of gradient of any scalar field is zero. This statement is most easily verified by writing it out in rectangular coordinates. If the scalar field is φ, then we can write:

$$\vec{\nabla} \times \vec{\nabla} \varphi = \begin{vmatrix} \hat{i} & \hat{j} & \hat{k} \\ \frac{\partial}{\partial x} & \frac{\partial}{\partial y} & \frac{\partial}{\partial z} \\ \frac{\partial \varphi}{\partial x} & \frac{\partial \varphi}{\partial y} & \frac{\partial \varphi}{\partial z} \end{vmatrix} = \hat{i} \left(\frac{\partial^2 \varphi}{\partial y \partial z} - \frac{\partial^2 \varphi}{\partial z \partial y} \right) + \cdots = 0 \tag{1.51}$$

This verifies the original statement. In operator notation:

$$\vec{\nabla} \times \vec{\nabla} = 0 \tag{1.52}$$

The divergence of any curl is also zero. This result is verified in rectangular coordinates by writing:

$$\vec{\nabla} \cdot \left(\vec{\nabla} \times \vec{F} \right) = \frac{\partial}{\partial x} \left(\frac{\partial F_x}{\partial y} - \frac{\partial F_y}{\partial z} \right) \qquad + \frac{\partial}{\partial y} \left(\frac{\partial F_x}{\partial z} - \frac{\partial F_z}{\partial x} \right) + \cdots = 0 \tag{1.53}$$

The two other possible second-order operations are taking the curl of the curl or the gradient of the divergence of a vector field. It is left as an exercise to show that in rectangular coordinates, the following is true as well:

$$\vec{\nabla} \times \left(\vec{\nabla} \times \vec{F} \right) = \vec{\nabla} \left(\vec{\nabla} \cdot \vec{F} \right) - \nabla^2 \vec{F} \tag{1.54}$$

Equation 1.54 indicates that the Laplacian of a vector is the vector whose rectangular components are the Laplacian of the rectangular components of the original vector. In any coordinate system other than rectangular, the Laplacian of a vector is defined by Equation 1.54.

The six possible combinations of differential operators and product are tabulated in Table 1.1, and they all can be verified easily in rectangular coordinate system.

A derivative of a product of more than two functions, or a higher than second-order derivative of a function, can be calculated by repeated applications of the identities in Table 1.1, which is therefore exhaustive. The formula can be easily remembered from the rules of vector algebra and ordinary differentiation.

Table 1.1 Differential vector identities

$\vec{\nabla} \cdot \vec{\nabla} \varphi = \nabla^2 \varphi$
$\vec{\nabla} \cdot \left(\vec{\nabla} \times \vec{F} \right) = 0$
$\vec{\nabla} \times \left(\vec{\nabla} \varphi \right) = 0$
$\vec{\nabla} \times \left(\vec{\nabla} \times \vec{F} \right) = \vec{\nabla} \left(\vec{\nabla} \cdot \vec{F} \right) - \nabla^2 \vec{F}$
$\vec{\nabla} (\varphi \psi) = \left(\vec{\nabla} \varphi \right) \psi + \varphi \vec{\nabla} \psi$
$\vec{\nabla} \left(\vec{F} \cdot \vec{G} \right) = \left(\vec{F} \cdot \vec{\nabla} \right) \vec{G} + \vec{F} \times \left(\vec{\nabla} \times \vec{G} \right) + \left(\vec{G} \cdot \vec{\nabla} \right) \vec{F} + \vec{G} \times \left(\vec{\nabla} \times \vec{F} \right)$
$\vec{\nabla} \cdot (\varphi \vec{F}) = \left(\vec{\nabla} \varphi \right) \cdot \vec{F} + \varphi \vec{\nabla} \cdot \vec{F}$
$\vec{\nabla} \cdot \left(\vec{F} \times \vec{G} \right) = \left(\vec{\nabla} \times \vec{F} \right) \cdot \vec{G} - \left(\vec{\nabla} \times \vec{G} \right) \cdot \vec{F}$
$\vec{\nabla} \times (\varphi \vec{F}) = \left(\vec{\nabla} \varphi \right) \times \vec{F} + \varphi \vec{\nabla} \times \vec{F}$
$\vec{\nabla} \times \left(\vec{F} \times \vec{G} \right) = \left(\vec{\nabla} \cdot \vec{G} \right) \vec{F} - \left(\vec{\nabla} \cdot \vec{F} \right) \vec{G} + \left(\vec{G} \cdot \vec{\nabla} \right) \vec{F} - \left(\vec{F} \cdot \vec{\nabla} \right) \vec{G}$

Some particular types of function come up often enough in electromagnetic theory that it is worth mentioning their various derivatives now.

For the function $\vec{F} = \vec{r}$, we can write the following relationship as:

$$\vec{\nabla} \cdot \vec{r} = 3 \quad \vec{\nabla} \times \vec{r} = 0 \left(\vec{G} \cdot \vec{\nabla} \right) \vec{r} = \vec{G} \nabla^2 \vec{r} = 0 \tag{1.55}$$

For a function that depends only on the distance $r = |\vec{r}| = \sqrt{x^2 + y^2 + z^2}$, we can write:

$$\varphi(r) \text{ or } \vec{F}(r) : \quad \vec{\nabla} = \frac{\vec{r}}{r} \frac{\mathrm{d}}{\mathrm{d}r} \tag{1.56}$$

For a function that depends on the scalar argument $\vec{A} \cdot \vec{r}$, where $\vec{A}$ is a constant vector:

$$\varphi \left(\vec{A} \cdot \vec{r} \right) \text{ or } \vec{F} \left(\vec{A} \cdot \vec{r} \right) : \quad \vec{\nabla} = \vec{A} \frac{\mathrm{d}}{\mathrm{d} \left(\vec{A} \cdot \vec{r} \right)} \tag{1.57}$$

For a function that depends on the argument $\vec{R} = \vec{r} - \vec{r}'$, where $\vec{r}'$ is treated as constant:

$$\vec{\nabla}_R = \hat{i} \frac{\partial}{\partial X} + \hat{j} \frac{\partial}{\partial Y} + \hat{k} \frac{\partial}{\partial Z} \tag{1.58}$$

where $\vec{R} = X\hat{i} + Y\hat{j} + Z\hat{k}$. If $\vec{r}$ is treated as constant instead:

$$\vec{\nabla} = -\vec{\nabla}\prime \tag{1.59}$$

where

$$\vec{\nabla}\prime = \hat{i}\,\frac{\partial}{\partial x\prime} + \hat{j}\,\frac{\partial}{\partial y\prime} + \hat{k}\,\frac{\partial}{\partial z\prime} \tag{1.60}$$

There are several possibilities for the extension of the divergence theorem and of Stokes's theorem. The most interesting of these is Green's theorem, which is:

$$\int_V \left(\psi\nabla^2\varphi - \varphi\nabla^2\psi \right)\mathrm{d}v = \oint_S \left(\psi\vec{\nabla}\varphi - \varphi\vec{\nabla}\psi \right)\cdot\hat{n}\,\mathrm{d}a \tag{1.61}$$

This theorem follows from the application of the divergence theorem to the vector:

$$\vec{F} = \psi\vec{\nabla}\varphi - \varphi\vec{\nabla}\psi \tag{1.62}$$

Using this vector $\vec{F}$ in the divergence theorem, we obtain:

$$\int_V \vec{\nabla}\cdot\left(\psi\nabla\varphi - \varphi\nabla\psi \right)\mathrm{d}v = \oint_S \left(\psi\vec{\nabla}\varphi - \varphi\vec{\nabla}\psi \right)\cdot\hat{n}\,\mathrm{d}a \tag{1.63}$$

Using the identity from Table 1.1 for the divergence of scalar times a vector gives:

$$\vec{\nabla}\cdot(\psi\nabla\varphi) - \vec{\nabla}\cdot(\varphi\nabla\psi) = \psi\nabla^2\varphi - \varphi\nabla^2\psi \tag{1.64}$$

Combining Equations 1.63 and 1.64 yields Green's theorem. Some other integral theorems are listed in Table 1.2.

This section is conclusion of our short course on vector analysis. Proof of many results is left to the reader as an exercise or extra study, and the approach just has been utilitarian; therefore, what we need to understand from the viewpoint of vector analysis have been developed to give us enough tools to go on with the rest of this book.

Table 1.2 Vector integral theorem

$$\int_S \hat{n}\times\vec{\nabla}\varphi\,\mathrm{d}a = \oint_C \varphi\,\mathrm{d}\vec{l}$$

$$\int_V \vec{\nabla}\varphi\,\mathrm{d}v = \oint_S \varphi\hat{n}\,\mathrm{d}a$$

$$\int_V \vec{\nabla}\times\vec{F}\,\mathrm{d}v = \oint_S \hat{n}\times\vec{F}\,\mathrm{d}a$$

$$\int_V \left(\vec{\nabla}\cdot\vec{G} + \vec{G}\cdot\vec{\nabla} \right)\vec{F}\,\mathrm{d}v = \oint_S \vec{F}\left(\vec{G}\cdot\hat{n} \right)\mathrm{d}a$$

1.4 Electrostatics

The subject of electricity is briefly touched upon for the rest of this chapter to provide us the fundamental of magnetism that we need in order to understand the science of plasma physics to go forward. We deal with the empirical concepts of charge and the force law between charges known as Coulomb's law. However, we use the mathematical tools of the previous section to express this law in other or more powerful formulations and then extended to the basic of plasma physics concept. The electric potential formulation and Gauss's law are very important to the subsequent development of the subject. Electric charge is a fundamental and characteristic property of the microscopic particles that makes up matter. In fact, all atoms are composed of photons, neutrons, and electrons, and two of these particles bear charges. However, even charge particles, the powerful electrical forces associated with these particles, are fairly well hidden in a macroscopic observation. The reason behind such statement exists because of the nature of existence of the two kinds of charges, namely, *positive* and *negative* charges, and an ordinary piece of matter contains approximately equal amounts of each kind.

It is understood from an experimental observation that charge can neither be created nor destroyed. The total charge of a closed system cannot change. From the macroscopic point of view, charges may be regrouped and combined in different ways; nevertheless, we may state that *net charge is conserved in a closed system* [1].

1.4.1 Coulomb's Law

To establish Coulomb's law, the three following statements can be summarized:

1. There are two and only two kinds of electric charge, now known as positive or negative.
2. Two point charges exert on each other forces that act along the line joining them and are inversely proportional to the square of the distance between them.
3. These forces are also proportional to the product of the charges, are repulsive for like charges, and are attractive for unlike charges.

The last two statements, with the first as preamble, all together, are known as Coulomb's law and for point charges may be concisely formulated in the vector notation as:

$$\vec{F}_1 = C_u \frac{q_1 q_2}{r_{12}^2} \frac{\vec{r}_{12}}{r_{12}}$$

$$\vec{r}_{12} = \vec{r}_1 - \vec{r}_2$$

$$(1.65a)$$

Fig. 1.6 Vector $\vec{r}_{12}$, extending between two points

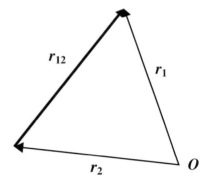

where $\vec{F}_1$ is the force on charge q_1, $\vec{r}_{12}$ is the vector to charge q_1 from charge q_2, r_{12} is the magnitude of vector $\vec{r}_{12}$, and C_u is a constant of proportionality about which is defined as to be equal to 1 in adoption with Gaussian system of units. Figure 1.6 will describe the vector $\vec{r}_{12}$ with respect to an arbitrary origin O.

In Fig. 1.6 vector $\vec{r}_{12}$ is extending from the point at the tip of vector $\vec{r}_2$ to the point at the tip of the vector $\vec{r}_1$ and clearly $\vec{r}_{12} = -\vec{r}_{21}$. Note that Coulomb's law applies to point charges, and in macroscopic sense, a "point charge" is one whose spatial dimensions are very small compared with any other length pertinent to the problem under consideration and that is why we use the term "point charge" in this sense.

In the MKS system, Coulomb's law for the force between two point charges can thus be written as:

$$\vec{F}_1 = \frac{1}{4\pi\varepsilon_0} \frac{q_1 q_2}{r_{12}^2} \frac{\vec{r}_{12}}{r_{12}} \tag{1.65b}$$

If more than two point charges are present, the mutual forces are determined by the repeated application of Equations 1.65a and 1.65b. In particular, if a system of N charges is considered, the force on the ith charge is given by:

$$\vec{F}_i = q_i \sum_{i \neq j}^{N} \frac{q_j}{4\pi\varepsilon_0 r_{ij}^3} \vec{r}_{ij} \tag{1.66}$$

$$\vec{r}_{ij} = \vec{r}_i - \vec{r}_j$$

where the summation on the right-hand side of Equation 1.66 is extended over all of the charges except the ith. Equation 1.66 is the superposition principle for forces, which says that the total force acting on a body is the vector sum of the individual forces that act on it. Note that in MKS unit, the value of Coulomb constant is: $C = 9 \times 10^9 \, \mathrm{N\,m^2/C^2}$.

There are cases such as fully ionized plasma that we may need to describe a charge distribution in terms of a *charge density function*; thus, it is defined as the limit of charge per unit volume as the volume becomes infinitesimal. However, care must be taken in applying this kind of description to atomic problems, since in such cases only small numbers of electrons are involved, and the process of taking the limit is meaningless. Nevertheless, aside from atomic case, we may proceed as though a segment of charges might be subdivided indefinitely; thus, we describe the charge distribution by means of point functions.

A *volume charge density* is defined by:

$$\rho = \lim_{\Delta V \to 0} \frac{\Delta q}{\Delta V} \qquad (1.67)$$

and a *surface charge density* is defined by:

$$\sigma = \lim_{\Delta S \to 0} \frac{\Delta q}{\Delta S} \qquad (1.68)$$

From the above statements and what has been said about point charge q, it is evident that ρ and σ are net charge, or excess charge, densities. It is worth to mention that in typical solid materials, even a very large charge density ρ will involve a change in the local electron density of only about one part 10^9.

Now that we have some concept of point charge and established Equations 1.65a, 1.65b, and 1.66, we extend our knowledge to a more general case. In this case, if the charge is distributed through a volume V with density ρ, and on the surface S that bounds the volume V with a surface density σ, then the force exerted by this charge distribution on a point charge q located at $\vec{r}$ is obtained from Equation 1.66 by replacing q_j with $\rho_j dv'_j$ or with $\sigma_j da'_j$ and processing to the limit as:

$$\begin{aligned}
\vec{F}_q &= \frac{q}{4\pi\varepsilon_0} \int_V \frac{\vec{r} - \vec{r}'}{\left|\vec{r} - \vec{r}'\right|^3} \rho(\vec{r}') dv' \\
&+ \frac{q}{4\pi\varepsilon_0} \int_S \frac{\vec{r} - \vec{r}'}{\left|\vec{r} - \vec{r}'\right|^3} \sigma(\vec{r}') da'
\end{aligned} \qquad (1.69)$$

The variable $\vec{r}'$ is used to locate a point within the charge distribution—that is, playing the role of the source point $\vec{r}_j$ in Equation 1.66 [1].

Equations 1.66 and 1.69 provide a ready means for obtaining an expression for the electric field due to a given distribution of charge as it is presented in Fig. 1.7 here, and electric field is discussed in the next section.

It may appear that the first integral in Equation 1.69 will diverge if point $\vec{r}$ should fall inside the charge distribution, but that is not the case at all.

In Fig. 1.7, the vector $\vec{r}$ defines the observation point (i.e., field point), and $\vec{r}'$ ranges over the entire charge distribution, including point charges.

Fig. 1.7 Geometry of $\vec{r}$, $\vec{r}'$, and $\vec{r} - \vec{r}'$

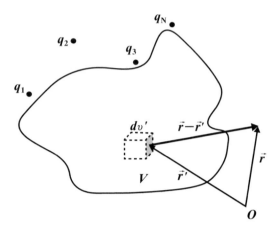

1.4.2   The Electric Field

Our first attempt to seek the electric field is for point charge for the sake of simplicity. The *electric field* at a point is defined operationally as the limit of the force on a test charge placed at the point to the charge of the test charge, and the limit being taken as the magnitude of the test charge goes to zero. The customary symbol for electric field in electromagnetic subject is $\vec{E}$ and no to be mistaken for energy presentation, which is the case by default. Thus, we can write:

$$\vec{E} = \lim_{q \to 0} \frac{\vec{F}_q}{q} \tag{1.70}$$

The limiting process is included in the definition of electric field to ensure that the test charge does not affect the charge distribution that produces $\vec{E}$.

Using Fig. 1.7, we let the charge distribution consist of N point q_1, q_2, $\cdots$, q_N located at the points $\vec{r}_1$, $\vec{r}_2, \cdots$, $\vec{r}_N$, respectively, and a volume distribution of charge specified by the charge density $\rho(\vec{r}')$ in the volume V and a surface distribution characterized by the surface charge density $\sigma(\vec{r}')$ on the surface S. If a test charge q is located at the point $\vec{r}$, it experiences force $\vec{F}$ given by the following equation due to the given charge distribution:

$$\vec{F} = \frac{q}{4\pi\varepsilon_0} \sum_{i=1}^{N} q_i \frac{\vec{r} - \vec{r}_i}{|\vec{r} - \vec{r}_i|^3}$$

$$+ \frac{q}{4\pi\varepsilon_0} \int_V \frac{\vec{r} - \vec{r}'}{|\vec{r} - \vec{r}'|^3} \rho(\vec{r}') \mathrm{d}v' \qquad (1.71)$$

$$+ \frac{q}{4\pi\varepsilon_0} \int_S \frac{\vec{r} - \vec{r}'}{|\vec{r} - \vec{r}'|^3} \sigma(\vec{r}') \mathrm{d}a'$$

In case of Equation 1.71, the electric field at the point $\vec{r}$ is then the limit of the ratio of this force to the test charge q. Since the ratio is independent of q, the electric field at $\vec{r}$ is just:

$$\vec{E}(\vec{r}) = \frac{1}{4\pi\varepsilon_0} \sum_{i=1}^{N} q_i \frac{\vec{r} - \vec{r}_i}{|\vec{r} - \vec{r}_i|^3}$$

$$+ \frac{1}{4\pi\varepsilon_0} \int_V \frac{\vec{r} - \vec{r}'}{|\vec{r} - \vec{r}'|^3} \rho(\vec{r}') \mathrm{d}v' \qquad (1.72)$$

$$+ \frac{1}{4\pi\varepsilon_0} \int_S \frac{\vec{r} - \vec{r}'}{|\vec{r} - \vec{r}'|^3} \sigma(\vec{r}') \mathrm{d}a'$$

Equation 1.72 is very general and in most cases, one or more of the terms will not be needed.

In order to complete the electromagnetic foundation circle, we also quickly note the general form of the potential energy associated with an arbitrary conservative force $\vec{F}(\vec{r}')$ as the following form:

$$U(\vec{r}) = - \int_{\text{ref.}\vec{r}} \vec{F}(\vec{r}') \cdot \mathrm{d}\vec{r}' \qquad (1.73)$$

where $U(\vec{r})$ is the potential energy at $\vec{r}$ relative to the reference point at which the potential energy is arbitrary taken to be zero. Proof is left to the reader by referring to the book of Reitz et al. [1].

1.4.3 Gauss's Law

One of the important relationships that exists between the integral of the normal component of the electric field over a closed surface and the total charge distribution enclosed by the surface is Gauss's law. To investigate that briefly here, we look

Fig. 1.8 An imaginary
closed surface S including
point charge at the origin

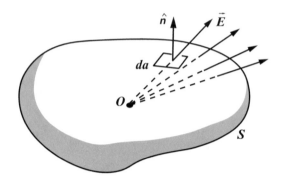

at the electric field $\vec{E}(\vec{r})$; for a point charge $\vec{r}q$ located at the origin, we can write the
following relation as before:

$$\vec{E}(\vec{r}) = \frac{q}{4\pi\varepsilon_0} \frac{\vec{r}}{r^3} \tag{1.74}$$

Consider the surface integral of the normal component of this electric field over a
closed surface such that shown in Fig. 1.8 here that encloses the origin and
consequently the charge q; then we can write:

$$\oint_S \vec{E} \cdot \hat{n}\,da = \frac{q}{4\pi\varepsilon_0} \oint_S \frac{\vec{r} \cdot \hat{n}}{r^3}\,da \tag{1.75}$$

The quantity $(\vec{r}/r) \cdot \hat{n}\,da$ is the projection of da on a plane perpendicular to $\vec{r}$. This
projected area divided by r^2 is the solid angle subtended by da, which is written in
$d\Omega$. It is clear from Fig. 1.9 that the solid angle subtended by the da is the same as
the solid angle subtended by da', an element of the surface area of the sphere S'
whose center is at the origin and whose radius is r'. It is then possible to write:

$$\oint_S \frac{\vec{r} \cdot \hat{n}}{r^3}\,da = \oint_{S'} \frac{\vec{r'} \cdot \hat{n}}{r'^3}\,da' = 4\pi \tag{1.76}$$

which shows the following equation in the spherical case described above:

$$\oint_S \vec{E} \cdot \hat{n}\,da = \frac{q}{4\pi\varepsilon_0}(4\pi) = \frac{q}{\varepsilon_0} \tag{1.77}$$

Figure 1.9 is illustrating the construction of the spherical surface S' as an aid to
evaluation of the solid angle subtended by da. If q lies outside of S, it is clear from
Fig. 1.10 that S can be divided into two areas, S_1 and S_2, each of which subtends the
same solid angle at the charge q. For S_2, however, the direction of the normal is
toward q, while for S_1 it is away from q.

Fig. 1.9 Construction of the spherical surface S'

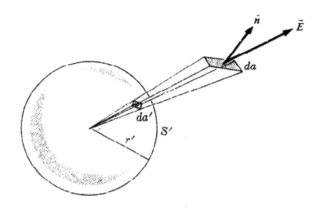

1. ANY CROSS PRODUCT

$$\vec{F} = q\vec{v} \times \vec{B} \qquad \vec{F} = I\vec{L} \times \vec{B}$$

$$\tau = \vec{r} \times \vec{F} \qquad \tau = \vec{\mu} \times \vec{B}$$

$$d\vec{B} = \frac{\mu_0 I}{4\pi} \frac{d\vec{s} \times \hat{r}}{r^2}$$

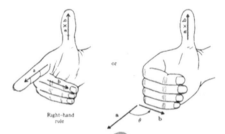

Right-hand rule

2. Direction of Magnetic Moment

Fingers: Current in Loop

Thumb: Magnetic Moment

3. Direction of Magnetic Field from Wire

Fingers: Magnetic Field

Thumb: Current

Fig. 1.10 Right-hand rule review

More details can be found in the reference by Reitz et al. [1], where readers need to go to; however, in case several point charges $q_1, q_2, \cdots, q_N$ are enclosed by the surface S, then the total electric field is given by the first term of Equation 1.72. Each charge subtends a full solid angle (4π); hence, Equation 1.77 becomes:

$$\oint_S \vec{E} \cdot \hat{n}\, da = \frac{1}{\varepsilon_0} \sum_{i=1}^{N} q_i \qquad (1.78)$$

The result in Equation 1.78 can be readily generalized to the case of a continuous distribution of charge characterized by a charge density [1].

1.5   Solution of Electrostatic Problems

Briefly we mention and write equations for the solution to an electrostatic problem which is straightforward for the case in which the charge distribution is everywhere specified, for then, as we have illustrated so far. The potential and electric fields are given as integral form over this charge distribution as:

$$\varphi(\vec{r}) = \frac{1}{4\pi\varepsilon_0} \int \frac{\mathrm{d}q'}{|\vec{r} - \vec{r}'|} \tag{1.79}$$

$$\vec{E}(\vec{r}) = \frac{1}{4\pi\varepsilon_0} \int \frac{(\vec{r} - \vec{r}')\mathrm{d}q'}{|\vec{r} - \vec{r}'|^3} \tag{1.80}$$

However, many of the problems that we encountered in real practice are not of this kind. If the charge distribution is not specified in advance, it may be necessary to determine the electric field *first*, before the charge distribution can be calculated.

1.5.1   Poisson's Equation

The only basic relationships we need here so far are developed in the preceding sections; thus, for that matter, we first write the differential form of Gauss's law as:

$$\vec{\nabla} \cdot \vec{E} = \frac{1}{\varepsilon_0} \rho \tag{1.81}$$

Equation 1.81 in a purely electrostatic field $\vec{E}$ may be expressed as minus the gradient of the potential φ:

$$\vec{E} = -\vec{\nabla}\varphi \tag{1.82}$$

Combining Equations 1.81 and 1.82, we obtain the following relation as:

$$\vec{\nabla} \cdot \vec{\nabla}\varphi = -\frac{\rho}{\varepsilon_0} \tag{1.83a}$$

Using vector identity as single differential operator as $\vec{\nabla} \cdot \vec{\nabla}$ or ∇^2, which is called the *Laplacian*, then we can express that the Laplacian is a scalar differential operator, and Equation 1.83a is a differential equation that is known as *Poisson's equation* and written as:

$$\nabla^2 \varphi = -\frac{\rho}{\varepsilon_0} \tag{1.83b}$$

The Laplace operator for Poisson's equation in rectangular, cylindrical, and spherical coordinate is presented here as well.

Rectangular or Cartesian coordinate:

$$\nabla^2 \varphi \equiv \frac{\partial^2 \varphi}{\partial x^2} + \frac{\partial^2 \varphi}{\partial y^2} + \frac{\partial^2 \varphi}{\partial z^2} = -\frac{\rho}{\varepsilon_0} \tag{1.84}$$

Cylindrical coordinate:

$$\nabla^2 \varphi \equiv \frac{1}{r}\frac{\partial}{\partial r}\left(r\frac{\partial \varphi}{\partial r}\right) + \frac{1}{r}\frac{\partial^2 \varphi}{\partial \theta} + \frac{\partial^2 \varphi}{\partial z^2} = -\frac{\rho}{\varepsilon_0} \tag{1.85}$$

Spherical coordinate:

$$\nabla^2 \varphi \equiv \frac{1}{r^2}\frac{\partial}{\partial r}\left(r^2\frac{\partial \varphi}{\partial r}\right) + \frac{1}{r^2 \sin\theta}\frac{\partial}{\partial \theta}\left(\sin\theta\frac{\partial \varphi}{\partial \theta}\right) + \frac{1}{r^2 \sin^2\theta}\frac{\partial^2 \varphi}{\partial \phi^2} = -\frac{\rho}{\varepsilon_0} \tag{1.86}$$

For the form of the Laplacian in other more complicated coordinated system, the reader is referred to the reference such as any vector analysis or advanced calculus books.

1.5.2   Laplace's Equation

Electrostatic problems are involving conductors; all the charges are either found on the surface of the conductors or in the form of fixed point charges. In these cases, charge density ρ is zero at most points in space, and in the absence of charge density, Poisson's equation reduces to the simpler form as follows:

$$\nabla^2 \varphi = 0 \tag{1.87}$$

Equation 1.87 is known as *Laplace's equation*.

1.6   Electrostatic Energy

From then on without further detail discussion and proof of different aspects of electrostatic equation, we just write them down as a basic knowledge, and we leave details to the readers to refer themselves to various subject books out in the open market.

Therefore, to go on with the subject in hand, we express that, under static condition, the entire energy of the charge system exists as potential energy, and in this section we are particularly concerned with the potential energy that arises from electrical interaction of the charges, the so-called *electrostatic energy U.*

We presented that the electrostatic energy U of a point charge is closely related to the electrostatic potential φ at the position of the point charge $\vec{r}$ as per Equation 1.73. In fact, if q is the magnitude of a particular point charge, then the work done by the force on the charge when it moves from position A to position B is given as:

$$\text{Work} = \int_A^B \vec{F} \cdot \mathrm{d}\vec{l} = q \int_A^B \vec{E} \cdot \mathrm{d}\vec{l}$$
$$= -q \int_A^B \vec{\nabla} \varphi \cdot \mathrm{d}\vec{l} = -q(\varphi_B - \varphi_A) \quad (1.88)$$

Here $\vec{F}$ has been assumed to be only the electric force $q\vec{E}$ at each point along the path or the *total work* is finalized to:

$$W = -q(\varphi_B - \varphi_A) \quad (1.89)$$

1.6.1   Potential Energy of a Group of Point Charges

The equation for potential energy of a group of point charges can be expressed as:

$$U = \sum_{j=1}^{m} W_j = \sum_{j=1}^{m} \left(\sum_{k=1}^{j-1} \frac{q_j q_k}{4\pi\varepsilon_0 r_{jk}} \right) \quad (1.90)$$

or in summary Equation 1.90 can be reduced to:

$$U = \frac{1}{2} \sum_{j=1}^{m} \sum_{k=1}^{m} \frac{q_j q_k}{4\pi\varepsilon_0 r_{jk}} \quad (1.91)$$

Note that on the second term of summation in Equation 1.91, where the prime is, the term $k = j$ specifically needs to be excluded, and Equation 1.91 may be written in a somewhat different way by noting that the final value of the potential φ at the jth point charge due to the other charges of the system is:

$$\varphi_j = \sum_{k=1}^{m} \frac{q_k}{4\pi\varepsilon_0 r_{jk}} \tag{1.92}$$

Thus, the electrostatic energy of the system is given as:

$$U = \frac{1}{2} \sum_{j=1}^{m} q_j \varphi_j \tag{1.93}$$

Proof of all the above equations is left to the readers.

1.6.2 Electrostatic Energy of a Charge Distribution

The electrostatic energy of an arbitrary charge distribution with volume density φ and surface density can be expressed based on assembled charge distribution by bringing in charge increments δq from a reference potential $\varphi_A = 0$. If the charge distribution is partly assembled and the potential at a particular point in the system is $\varphi'(x, y, z)$, then, from Equation 1.89, the work required to place δq at this point is written as:

$$\delta W = \varphi'(x, y, z)\delta q \tag{1.94}$$

In this equation the charge increment δq may be added to a volume element located at (x, y, z), so that $\delta q = \delta\rho\Delta v$, or may be added to a surface element at the point in question, in which case $\delta q = \delta\rho\Delta a$. The total electrostatic energy of the assembled charge distribution is obtained by summing contributions of Equation 1.94.

Let us assume at any stage of the charging process, all charge densities will be at the same fraction of their final values and represented by the symbol α, and if the final values of the charge densities are given by the function $\varphi(x, y, z)$ and $\sigma(x, y, z)$, then the charge densities at an arbitrary stage are $\alpha\varphi(x, y, z)$ and $\alpha\sigma(x, y, z)$. Furthermore, if the increments in these densities are $\delta\rho = \varphi(x, y, z)d\alpha$ and $\delta\sigma = \sigma(x, y, z)d\alpha$, then the total electrostatic energy, which is obtained by summing Equation 1.94, is given by:

$$U = \int_0^1 \delta d \int_V \varphi(x, y, z)\varphi'(x, y, z)dv$$
$$+ \int_0^1 \delta d \int_S \sigma(x, y, z)\varphi'(x, y, z)da \tag{1.95}$$

However, since all charges are at the same fraction, α is readily done and yields as:

$$U = \frac{1}{2} \int_V \rho(\vec{r})\varphi(\vec{r})\mathrm{d}\upsilon + \frac{1}{2} \int_s \sigma(\vec{r})\varphi(\vec{r})\mathrm{d}a \qquad (1.96)$$

This equation provides the desired result for the energy of a charge distribution. If all space is filled with a single dielectric medium except for certain conductors, the potential is then given by:

$$\varphi(\vec{r}) = \frac{1}{4\pi\varepsilon} \int_V \frac{\varphi(\vec{r}')\mathrm{d}\upsilon'}{|\vec{r} - \vec{r}'|} + \frac{1}{4\pi\varepsilon} \int_V \frac{\sigma(\vec{r}')\mathrm{d}a'}{|\vec{r} - \vec{r}'|} \qquad (1.97)$$

Equations 1.96 and 1.97 are the generalization of Equations 1.92 and 1.93 for point charges. The latter can be recovered as a special case letting the following relationships as:

$$\begin{aligned}
\rho(\vec{r}) &= \sum_{j=1}^{m} q_j \delta(\vec{r} - \vec{r}_j) \\
\rho(\vec{r}') &= \sum_{k=1}^{m}{}' q_k \delta(\vec{r} - \vec{r}_k)
\end{aligned} \qquad (1.98)$$

where again, the prime on the second summation in Equation 1.98 is an indication of the term $k = j$ which is excluded when the double sum is constructed. Note that when ρ is a continuous distribution, the vanishing of the denominator in Equation 1.97 does not cause the integral to diverge, and it is unnecessary to exclude the point $\vec{r}' = \vec{r}$.

The last integral involves, in part, integration over the surface of the conductor of interest; however, since a conductor is an equipotential region, each of these integrations may be done as:

$$\frac{1}{2} \int_{\text{conductor} j} \sigma\varphi\mathrm{d}a = \frac{1}{2} Q_j \varphi_j \qquad (1.99)$$

where Q_j is the charge on the jth conductor.

Equation 1.96 for *electrostatic energy of a charge distribution*, which includes a conductor, then becomes:

$$U = \frac{1}{2} \int_V \rho\varphi\mathrm{d}\upsilon + \frac{1}{2} \int_{S'} \sigma\varphi\mathrm{d}a + \frac{1}{2} \sum_j Q_j \varphi_j \qquad (1.100)$$

where in Equation 1.100, the last summation is over all conductors and the surface integral is restricted to nonconducting surfaces.

Furthermore, in many practical problems of interest, all of the charges reside on the surfaces of the conductor. In these circumstances Equation 1.100 reduces to the following form as:

$$U = \frac{1}{2} \sum_j Q_j \varphi_j \qquad (1.101)$$

Equation 1.101 is derived based on starting with uncharged macroscopic conductors that were gradually charged by bringing in charge increments. Thus, the energy is described by Equation 1.101 including both interaction energy between different conductors and the self-energies of the charge on each individual conductor.

1.6.3   Forces and Torques

Thus far, we have developed to some extent a number of alternative procedures for calculating the electrostatic energy of a charge system. We now take an attempt to establish the force on one of the objects in the charge system which may be calculated from knowledge of this electrostatic energy.

If we dealing with an isolated system composed of conductors, point charges, and dielectrics and we all one of these items to make a small displacement $d\vec{r}$ under the influence of the electrical force $\vec{F}$ acting upon it. The work performed by the electrical force on the system in these circumstances is:

$$dW = \vec{F} \cdot d\vec{r} = F_x dx + F_y dy + F_z dz \qquad (1.102)$$

Since we assume the system is isolated, this work is done at the expense of the electrostatic energy U. In other words, according to Equation 1.88, we can write:

$$dW = -dU \qquad (1.103)$$

Combining Equations 1.102 and 1.103, the result is:

$$-dU = F_x dx + F_y dy + F_z dz \qquad (1.104)$$

and

$$
\begin{aligned}
F_x &= -\frac{\partial U}{\partial x} \\
F_y &= -\frac{\partial U}{\partial y} \\
F_z &= -\frac{\partial U}{\partial z}
\end{aligned}
\qquad (1.105)
$$

Therefore, sets of Equation 1.105 indicate that $\vec{F}$ is a conservative force and $\vec{F} = -\vec{\nabla} U$. If the object under consideration is constrained to move in such a way that it rotates about an axis, then Equation 1.102 may be replaced by the following equation as:

$$dW = \vec{\tau} \cdot d\vec{\theta} \qquad (1.106)$$

where $\vec{\tau}$ is the electrical torque and $d\vec{\theta}$ is the differential angular displacement. Writing $\vec{\tau}$ and $d\vec{\theta}$ in terms of their components, $(\tau_1, \ \tau_2, \tau_2)$ and $(d\theta_1, d\theta_2, d\theta_3)$, and combining Equations 1.103 and 1.106, we obtain the following relationships:

$$
\begin{aligned}
\tau_1 &= -\frac{\partial U}{\partial \theta_1} \\
\tau_2 &= -\frac{\partial U}{\partial \theta_2} \\
\tau_3 &= -\frac{\partial U}{\partial \theta_3}
\end{aligned}
\qquad (1.107)
$$

This proves that our goal has been achieved and we can write:

$$
\begin{cases}
F_x = -\left(\frac{\partial U}{\partial x}\right)_Q \\
\tau_1 = -\left(\frac{\partial U}{\partial \theta_1}\right)_Q
\end{cases}
\qquad (1.108a)
$$

$$
\begin{cases}
F_y = -\left(\frac{\partial U}{\partial y}\right)_Q \\
\tau_2 = -\left(\frac{\partial U}{\partial \theta_2}\right)_Q
\end{cases}
\qquad (1.108b)
$$

$$
\begin{cases}
F_z = -\left(\frac{\partial U}{\partial x}\right)_Q \\
\tau_3 = -\left(\frac{\partial U}{\partial \theta_3}\right)_Q
\end{cases}
\qquad (1.108c)
$$

where the subscript Q has been added to denote that the system is isolated, and hence its total charge remains constant during the displacement $d\vec{r}$ or $d\vec{\theta}$.

Now, we are at the stage that we need to talk electromagnetic force that is known as Lorentz force here.

The electromagnetic field exerts the following force (often called the Lorentz force) on charged particles:

$$\vec{F} = q\vec{E} + q\vec{v} \times \vec{B} \qquad (1.109)$$

where vector $\vec{F}$ is the force that a particle with charge q experiences, $\vec{E}$ is the electric field at the location of the particle, v is the velocity of the particle, and $\vec{B}$ is the magnetic field at the location of the particle.

The above equation illustrates that the Lorentz force is the sum of two vectors. One is the cross product of the velocity and magnetic field vectors. Based on the properties of the cross product, this produces a vector that is perpendicular to both the velocity and magnetic field vectors. The other vector is in the same direction as the electric field. The sum of these two vectors is the Lorentz force.

Therefore, in the absence of a magnetic field, the force is in the direction of the electric field, and the magnitude of the force is dependent on the value of the charge and the intensity of the electric field. In the absence of an electric field, the force is perpendicular to the velocity of the particle and the direction of the magnetic field. If both electric and magnetic fields are present, the Lorentz force is the sum of both of these vectors.

Therefore, in summary we can express that the classical theory of electrodynamics is built upon Maxwell's equations and the concepts of electromagnetic field, force, energy, and momentum, which are intimately tied together by Poynting's theorem and the Lorentz force law. Whereas Maxwell's macroscopic equations relate the electric and magnetic fields to their material sources (i.e., charge, current, polarization, and magnetization), Poynting's theorem governs the flow of electromagnetic energy and its exchange between fields and material media, while the Lorentz law regulates the back-and-forth transfer of momentum between the media and the fields. As it turns out, an alternative force law, first proposed in 1908 by Einstein and Laub, exists that is consistent with Maxwell's macroscopic equations and complies with the conservation laws as well as with the requirements of special relativity. While the Lorentz law requires the introduction of hidden energy and hidden momentum in situations where an electric field acts on a magnetic material, the Einstein-Laub formulation of electromagnetic force and torque does not invoke hidden entities under such circumstances. Moreover, the total force and the total torque exerted by electromagnetic fields on any given object turn out to be independent of whether force and torque densities are evaluated using the Lorentz law or in accordance with the Einstein-Laub formulas. Hidden entities aside, the two formulations differ only in their predicted force and torque distributions throughout the material media. Such differences in distribution are occasionally measurable and could serve as a guide in deciding which formulation, if either, corresponds to physical reality.

Furthermore, to have some general idea about Poynting's theorem, we can say that, in electrodynamics, Poynting's theorem is a statement of conservation of energy for the electromagnetic field. Moreover, it is in the form of a partial differential equation, due to the British physicist John Henry Poynting. Poynting's theorem is analogous to the work-energy theorem in classical mechanics, and mathematically similar to the continuity equation, because it relates the energy stored in the electromagnetic field to the work done on a charge distribution (i.e., an electrically charged object), through energy flux. A detail of deriving this theorem is beyond the scope of this book and we leave to the readers to refer to some other classical electrodynamics books.

However, in general we can say that this theorem is an energy balance and the following statement does apply:

The rate of energy transfer (per unit volume) from a region of space equals the rate of work done on a charge distribution plus the energy flux leaving that region.

A second statement can also explain the theorem—"The decrease in the electromagnetic energy per unit time in a certain volume is equal to the sum of work done by the field forces and the net outward flux per unit time."

Mathematically, the above statement can be expressed and is summarized in differential form as below:

$$-\frac{\partial u}{\partial t} = \vec{\nabla} \cdot \vec{S} + \vec{J} \cdot \vec{E} \tag{1.110}$$

where $\vec{\nabla} \cdot \vec{S}$ is the divergence of the Poynting vector or energy flow and $\vec{J} \cdot \vec{E}$ is the rate at which the fields do work on a charged object, ($\vec{J}_{\rm f}$ is the free current density corresponding to the motion of charge, and $\vec{E}$ is the electric field and • is the dot product). The energy density u is given by:

$$u = \frac{1}{2}\left(\vec{E} \cdot \vec{D} + \vec{B} \cdot \vec{H}\right) \tag{1.111}$$

In this equation $\vec{D}$ is the electric displacement field, $\vec{B}$ is the magnetic flux density, and $\vec{H}$ is the magnetic field strength. Since only some of the charges are free to move, and $\vec{D}$ and $\vec{H}$ fields exclude the "bound" charges and currents in the charge distribution (by their definition), one obtains the free current density $\vec{J}_{\rm f}$ in Poynting's theorem, rather than the total current density $\vec{J}$.

The integral form of Poynting's theorem can be established via utilization of divergence theorem expressed before as:

$$-\frac{\partial}{\partial t}\int_V u\mathrm{d}V = \oiint_{\partial V} \vec{S} \cdot \mathrm{d}\vec{A} + \int_V \vec{J} \cdot \vec{E}\mathrm{d}V \tag{1.112}$$

where ∂V is the boundary of volume V and the shape of the volume is arbitrary, but fixed for the calculation.

In summary of all past couple section in this chapter we can in perspectives that are presented by Fig. 1.10, below

1.7   Maxwell's Equations

In order to understand the physics of plasma and associated subject such as magnetohydrodynamic equations that are known as MHD in particular encountering confinement of plasma as a way of driving fusion energy, we need to have some understanding of the sets of equations that are known as Maxwell's equations.

We are at the point and ready to introduce the keynote of Maxwell's electromagnetic theory as a brief course and what is so-called displacement current. We shall now write all classical, i.e., non-quantum, electromagnetic phenomena that are governed by *Maxwell's equations*, which take the form as follows:

$$\vec{\nabla} \cdot \vec{E} = \frac{\rho}{\varepsilon_0} \text{ Also known as Coulomb's Law} \tag{1.113}$$

$$\vec{\nabla} \cdot \vec{B} = 0 \text{ Also known as Gauss's Law} \tag{1.114}$$

$$\vec{\nabla} \times \vec{E} = -\frac{\partial \vec{B}}{\partial t} \text{ Also known as Faraday's Law} \tag{1.115}$$

$$\vec{\nabla} \times \vec{B} = \mu_0 \vec{J} + \mu_0 \varepsilon_0 \frac{\partial \vec{E}}{\partial t} \text{ Also known as Ampere's Law} \tag{1.116}$$

All the quantities in the above equations are defined as before. Here, $\vec{E}(\vec{r}, t)$, $\vec{B}(\vec{r}, t)$, $\rho(\vec{r}, t)$, and $\vec{J}(\vec{r}, t)$ represent the *electric field strength, magnetic field strength, electric charge density*, and *electric current density*, respectively. Moreover, $\varepsilon_0 = 8.8542 \times 10^{-2} \, \text{C}^2\text{N}^{-1}\text{m}^{-2}$ is the *electric permittivity of free space*, whereas $\mu_0 = 4\pi \times 10^{-7} \, \text{N A}^{-2}$ is the *magnetic permeability of free space*. As is well known, Equation 1.113 is equivalent to *Coulomb's law* for the electric fields generated by point charges. Equation 1.114 is equivalent to the statement that magnetic monopoles do not exist, which implies that magnetic field lines can never begin or end. Equation 1.115 is equivalent to *Faraday's law of electromagnetic induction*. Finally, Equation 1.116 is equivalent to Biot-Savart's law for the magnetic fields generated by line currents and augmented by the induction of magnetic fields by changing electric fields.

Maxwell's equations are linear in nature. In other words, if $\rho \rightarrow \alpha\rho$ and $\vec{J} \rightarrow \alpha\vec{J}$, where α is an arbitrary spatial and temporal constant, then it is clear from Equations 1.113 to 1.116 that $\vec{E} \rightarrow \alpha\vec{E}$ and $\vec{B} \rightarrow \alpha\vec{B}$. The linearity of Maxwell's equations account for the well-known fact that the electric fields generated by point charges as well as the magnetic fields generated by line currents are superposable.

Taking the divergence of Equation 1.113, and combining the resulting expression with Equation 1.113, we obtain:

$$\frac{\partial \rho}{\partial t} + \vec{\nabla} \cdot \vec{J} = 0 \tag{1.117}$$

In integral form, making use of the divergence theorem, this equation becomes:

$$\frac{d}{dt} \int_V \rho \, dV + \int_S \vec{J} \cdot d\vec{S} = 0 \tag{1.118}$$

where V is a fixed volume bounded by a surface S. The volume integral represents the net electric charge contained within the volume, whereas the surface integral represents the outward flux of charge across the bounding surface. The previous equation, which states that the net rate of change of the charge contained within the volume V is equal to minus the net flux of charge across the bounding surface S, is clearly a statement of the *conservation of electric charge*. Thus, Equation 1.117 is the differential form of this conservation equation.

As is well known, a point electric q moving with velocity $\vec{v}$ in the presence of an electric field $\vec{E}$ and a magnetic field $\vec{B}$ experiences a force that is known as Lorentz force and was expressed by Equation 1.109 as before. Likewise, distribution of charge density ρ and current density $\vec{J}$ experiences a force density that is given as:

$$\vec{f} = \rho \vec{E} + \vec{J} \times \vec{B} \tag{1.119}$$

This is the extent of our presentation for Maxwell's equations within this book; further deviation of these equations can be found in any classical electrodynamics books out there [1].

1.8   Debye Length

Debye length is an important aspect of plasma physics, and it is a quantity which is a measure of the shielding distance or thickness of the charged particle cloud also called sheath in plasma. One of the most significant properties of plasma is its tendency to maintain electrically neutral—that is, its tendency to balance positive (ion) and negative (electron) space charge in each macroscopic volume element. A slight imbalance in the space charge densities gives rise to strong electrostatic forces that act, wherever possible, in the direction of restoring neutrality. On the other hand, if plasma is deliberately subjected to an external electric field, the space charge densities will adjust themselves so that the major part of the plasma is shielded from the field.

To carry out this subject further, we can pay our attention to Poisson's equation and seek a solution for that equation in case of a point charge $+Q$ that is introduced into a plasma and thereby subjecting the plasma to an electric field for simplicity of analyses. Under these conditions, negative electrons existing in plasma find it energetically tendency to move closer to this positive charge favorably, whereas positive ions tend to move away from it. Under equilibrium conditions, the probability of finding a charged particle in a particular region of potential energy U is proportional to the Boltzmann factor as $\exp(-U/kT)$. Thus, the electron density n_e is given by the following equation as:

$$n_e = n_0 \exp\left(e \frac{(\varphi - \varphi_0)}{kT}\right) \qquad (1.120)$$

For Equation 1.120, the following quantities are in order and they are:

φ = the local potential

φ_0 = the reference potential or in our case plasma potential

T = the absolute temperature of the plasma

k = the Boltzmann constant

n_0 = the electron density in regions where $\varphi = \varphi_0$

If n_0 is also the positive ion density in regions of potential φ_0, then positive ion density n_i is also given by the similar relation as Equation 1.120 and that is:

$$n_i = n_0 \exp\left(-e \frac{(\varphi - \varphi_0)}{kT}\right) \qquad (1.121)$$

Now that we have set up the initial conditions, first we attempt to derive Debye length by means of Poisson's equation and then show its use in plasma physics and as criteria to identify a definition that plasmas fall into it.

A particular solution of Poisson's equation for potential φ is carried out here, from one-dimensional spherical symmetry around the radius coordinate of r, and we start with the following differential equation as:

$$\frac{1}{r^2}\frac{d}{dr}\left(r^2 \frac{d\varphi}{dr}\right) = -\frac{1}{\varepsilon_0}(n_i e - n_e e) = \frac{2n_0 e}{\varepsilon_0}\sinh\left(e \frac{(\varphi - \varphi_0)}{kT}\right) \qquad (1.122)$$

The differential Equation 1.122 is nonlinear and hence must be integrated numerically. On the other hand, an approximate solution to Equation 1.122, which is rigorous at high-temperature plasma, is adequate for these purposes here. If $kT > e\varphi$, then $\sinh(e\varphi/kT) = e\varphi/kT$, and the differential Equation 1.122 reduces to the following and simple form as:

$$\frac{1}{r^2}\frac{d}{dr}\left(r^2 \frac{d\varphi}{dr}\right) = \frac{2n_0 e^2}{\varepsilon_0 kT}(\varphi - \varphi_0) \qquad (1.123)$$

The solution to this equation is found to be (readers can carry out the solution; as hint use Taylor series expansion for $|e\varphi/kT| \ll 1$ to drop the second order and higher terms off in expansion of $e\varphi/kT + \frac{1}{2}(e\varphi/kT)^2 + \cdots$):

$$\varphi = \varphi_0 + \frac{Q}{4\pi\varepsilon_0 r}\exp\left(-\frac{r}{h}\right) \qquad (1.124)$$

where r is the distance from the point charge $+Q$ and λ_D, the Debye shielding distance or Debye length, is given by:

$$\lambda_D = \left(\frac{\varepsilon_0 kT}{2n_0 e^2} \right) \tag{1.125}$$

Thus, the redistribution of electrons and ions in the gas is such as to screen out $+Q$ completely in a distance of a few λ_D.

The quantity λ_D as we have said before is called the Debye length and is the measure of the shielding or thickness of the charge cloud, which is also known as sheath. Note that as the density increases, λ_D decreases, as one would expect, since each layer of plasma contains more electrons. In addition, λ_D increases with increasing kT. Without thermal agitation, the charge cloud would collapse to an infinitely thin layer. Last but not least, it is the *electron* temperature which is used in the definition of λ_D and that is $T = T_e$, because the electrons are being more mobile than their counterpart ions. In general shielding do the moving so as to create a surplus or deficit of negative charge. Only in special situations is this not true. The following are set of useful forms of Equation 1.125 and they are as follows:

$$\lambda_D = 69(T_e/n)^{1/2} m \quad T_e \text{ in } {}^0K \tag{1.126a}$$

$$\lambda_D = 7430(T_e/n)^{1/2} m \quad kT_e \text{ in eV} \tag{1.126b}$$

1.9 Physics of Plasmas

An ionized gas is called a plasma if the Debye length, λ_D, is small compared with other physical dimensions of interest. This restriction is not severe so long as ionization of the gas is appreciable. Other conditions that will make an ionized gas fall in the category of plasma can be described in the following statements.

One criterion for an ionized gas to be called plasma is that it needs to be dense enough that λ_D is much smaller than the dimension L of a system, and if this dimension is much larger than λ_D, in other words $\lambda_D \ll L$, then local concentrations of charge arise or external potentials are introduced into the system. The system could be a magnetron or klystron.

The phenomenon of Debye shielding also occurs—in modified form—in single-species systems, such as the electron streams in klystrons and magnetrons or the proton beams in a cyclotron. Under these situations, any local bunching of particles causes a large unshielded electric field unless the density is extremely low, which is more often is the case.

The Debye shielding picture that we have painted above is valid only if there are enough particles in charge cloud or sheath. Thus, it is clear that if there is only one or two particles in the sheath region, Debye shielding would not be a statistically valid concept from the viewpoint of electromagnetic physics. Using Equation 1.120 in a general form, we can compute the number of N_D particles in a Debye sphere as:

$$N_D = n\frac{4}{3}\pi\lambda_D^3 = 1.38 \times 10^6 T^{3/2}/n^{1/2} \quad T \text{ in } K \qquad (1.127)$$

In addition to $\lambda_D \ll L$, "collective behavior" requires [2]:

$$N_D \gg 1 \qquad (1.128)$$

Furthermore, to qualify an ionized gas as plasma, we can define more criteria. The two conditions above were given that an ionized gas must satisfy to be a plasma. A third condition has to do with collisions. The ionized gas in an airplane's jet exhaust, for example, does not qualify as a plasma because the charged particles collide so frequently with neutral atoms that their motion is controlled by ordinary hydrodynamic forces rather than by electromagnetic forces [2].

If ω is the frequency of typical plasma oscillations and τ is the mean time between collisions with neutral atoms, we require $\omega\tau > 1$ for the gas to behave like plasma rather than a neutral gas. Therefore, the three conditions a plasma must satisfy are therefore:

1.
$$\lambda_D \ll L$$

2.
$$N_D \gg 1$$

3.
$$\omega\tau > 1$$

As you can see, the above three conditions are necessary for an ionized gas to be called plasma

1.10 Fluid Description of Plasma

Before paying our attention and departing for the actual derivation of the magnetohydrodynamics (MHD) equation, which is the topic of our next section in this chapter, it is helpful to discuss briefly some general concepts of fluid dynamics.

Fluid equations are probably the most widely used equations for the description of inhomogeneous plasmas. While the phase fluid, which is governed by the Boltzmann equation, represents the first example, many applications do not require the precise velocity distribution at any point in space.

Ordinary fluid equations for gases and plasmas can be obtained from the Boltzmann equation or can be derived using properties like the conservation of mass, momentum, and energy of the fluid. For the following chapter, we will derive a single set of ordinary fluid equations for a plasma and examine properties such an equilibrium and waves for these equations.

To further investigate the fluid aspect of plasma, we look at the equations of kinetic theory and taking a fundamental equation such as $f(\vec{r}, \vec{v}, t)$ under consideration, which satisfies the Boltzmann equation as follows:

$$\frac{\partial f(\vec{r}, \vec{v}, t)}{\partial t} + \vec{v} \cdot \vec{\nabla} f(\vec{r}, \vec{v}, t) + \frac{\vec{F}}{m} \cdot \frac{\partial f(\vec{r}, \vec{v}, t)}{\partial \vec{v}} = \left(\frac{\partial f(\vec{r}, \vec{v}, t)}{\partial t} \right)_c \qquad (1.129)$$

In Equation 1.129, $\vec{F}$ is the force acting on the particles, and $(\partial f(\vec{r}, \vec{v}, t)/\partial t)_c$ is the time rate of change of $f(\vec{r}, \vec{v}, t)$ due to collisions. The symbol $\vec{\nabla}$, as usual, is for the gradient in (x, y, z) space. The symbol $\partial/\partial \vec{v}$ or $\vec{\nabla}_{\vec{v}}$ stands for the gradient in velocity space and it is written as:

$$\frac{\partial}{\partial \vec{v}} = \hat{x} \frac{\partial}{\partial v_x} + \hat{y} \frac{\partial}{\partial v_y} + \hat{z} \frac{\partial}{\partial v_z} \qquad (1.130)$$

The Boltzmann equation becomes more meaningful if one should remember that function $f(\vec{r}, \vec{v}, t)$ is a function of seven independent variables, which include three for space (x, y, z), three for components of velocity (v_x, v_y, v_z), and the seventh one that accounts for time t; therefore, we can expand Equation 1.129 to all its seven variables and write down:

$$\frac{df}{dt} = \frac{\partial f}{\partial t} + \frac{\partial f}{\partial x} \frac{dx}{dt} + \frac{\partial f}{\partial y} \frac{dy}{dt} + \frac{\partial f}{\partial z} \frac{dz}{dt} + \frac{\partial f}{\partial v_x} \frac{dv_x}{dt} + \frac{\partial f}{\partial v_y} \frac{dv_y}{dt} + \frac{\partial f}{\partial v_z} \frac{dv_z}{dt} \qquad (1.131)$$

Here, $\partial f/\partial t$ is the *explicit* dependence on time. The next three terms are just $\vec{v} \cdot \vec{\nabla} f(\vec{r}, \vec{v}, t)$. With the help of Newton's third law, we can write:

$$m \frac{d\vec{v}}{dt} = \vec{F} \qquad (1.132)$$

As it can be seen from Equation 1.132, the last three terms are recognized as $(\vec{F}/m) \cdot (\partial f/\partial \vec{v})$.

Additionally, the total derivative term presented by df/dt can be interpreted as the rate of change as seen in a frame moving with the particles. However, here we need to be concerned with particles to be moving in six-dimensional space $(\vec{r}, \vec{v})$, i.e., three in (x, y, z) direction and the associate three components of velocity (v_x, v_y, v_z) in their corresponding directions as well.

df/dt is the convective derivative in phase space and the Boltzmann equation simply says that df/dt is zero, unless there are collisions. This should be true and can be seen from the one-dimensional example shown in Fig. 1.11 here. Figure 1.11 illustrates a group of points in phase space, representing the position and velocity coordinates of a group of particles, and retains the same phase-space density as it moves with time.

Fig. 1.11 Illustration of group points in phase space

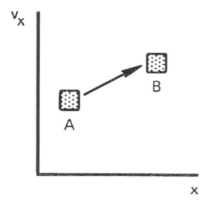

Taking Fig. 1.11 under consideration and assuming the group of particles in an infinitesimal element $dx\,dv_x$ at point A that all have velocity v_x and position x, then the density of particles in this phase space is just $f(x, v_x)$. As the time passes, these particles will move to a different position in x because of their velocity v_x and will change their velocity due to the result of the force acting on them.

Since the forces depend on x and v_x only, all the particles at A will be accelerated in the same amount. After a time t, all the particles that will arrive at B will be the same as at A. If there exist any collusions, then the particles can be scattered and $f(\vec{r}, \vec{v}, t)$ can be changed by the term $(\partial f(\vec{r}, \vec{v}, t)/\partial t)_c$. In sufficiently hot plasma, collision can be neglected, and furthermore, if the force $\vec{F}$ is entirely electromagnetic, Equation 1.129 takes the speed form:

$$\boxed{\frac{\partial f}{\partial t} + \vec{v} \cdot \vec{\nabla} f + \frac{q}{m}\left(\vec{E} + \vec{v} \times \vec{B}\right) \cdot \frac{\partial f}{\partial \vec{v}} = 0} \qquad (1.133)$$

Equation 1.133 is representing the Vlasov equation, and because of its comparative simplicity, this is the equation that is most commonly studied in kinetic theory. If there exist collisions with neutral atoms, then the collision term in Equation 1.129 can be approximated to:

$$\left(\frac{\partial f(\vec{r}, \vec{v}, t)}{\partial t}\right)_c = \frac{f_n(\vec{r}, \vec{v}, t) - f(\vec{r}, \vec{v}, t)}{\tau} \qquad (1.134)$$

where $f_n(\vec{r}, \vec{v}, t)$ is the distribution function of the neutral atoms, and τ is a constant collision time. This equation is called *Krook collision term*.

The fluid equation of motion including collisions for any species is given by the following relation:

Fig. 1.12 Illustration of the
definition of cross section

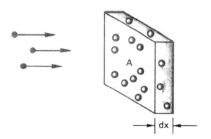

$$mn\frac{d\vec{v}}{dt} = mn\left[\frac{\partial \vec{v}}{\partial t} + \left(\vec{v} \cdot \vec{\nabla}\right)\vec{v}\right] = \pm en\vec{E} - \nabla\rho - mnv\,\vec{v} \qquad (1.135)$$

where the sign $\pm$ is indication of the sign of the charge and v is generally called the
collision frequency of plasma particles and is written as $v = n_n \overline{\sigma v}$, with σ being the
cross-sectional area and v is the particle velocity in a Maxwellian distribution and
n_n is number neutral atoms per m^3 in slab of area A and thickness dx as illustrated in
Fig. 1.12 here.

It is the kinetic generalization of the collision term in Equation 1.135. When
there are Coulomb collisions, Equation 1.129 can be approximated by:

$$\frac{df}{dt} = -\frac{\partial}{\partial \vec{v}} \cdot (f\langle\nabla \vec{v}\rangle)\frac{1}{2}\frac{\partial^2}{\partial \vec{v}\partial \vec{v}} : (f\langle\nabla \vec{v}\nabla \vec{v}\rangle) \qquad (1.136)$$

Equation 1.136 is called the Fokker-Plank equation and it takes into account binary
Coulomb collisions only [1].

1.11 MHD

Magnetohydrodynamics (MHD) describes the "slow" evolution of an electrically
conducting fluid—most often a plasma consisting of electrons and protons (perhaps
seasoned sparingly with other positive ions). In MHD, "slow" means evolution on
time scales longer than those on which individual particles are important or on
which the electrons and ions might evolve independently of one another. Briefly we
can say that MHD falls in the following descriptions as:

- MHD stands for magnetohydrodynamics.
- MHD is a simple, self-consistent fluid description of a fusion plasma.
- Its main application involves the macroscopic equilibrium and stability of a
  plasma.

Basically MHD can be described as coupling of fluid dynamics equations with
Maxwell's equations which results in MHD equations, and together these sets of
equation are often used to describe the equilibrium state of the plasma. MHD can

also be used to derive plasma waves, but it is considerably less accurate than the two-fluid equations we are familiar with and have used in our fluid mechanics knowledge.

Moreover, to define the plasma equilibrium and stability, we can categorize the definition into the following format as well and they are:

- Why separate the macroscopic behavior into two pieces?
- Even though MHD is simple, it still involves nonlinear 3D + time equations.
- This is tough to solve.
- Separation simplifies the problem.
- Equilibrium requires 2D nonlinear time independent.
- Stability requires 3D + time, but is linear.
- This enormously simplifies the analysis.

We need to understand the idea behind the plasma equilibrium, so it allows in case of Magnetic Confinement Fusion (MCF) to design a magnet system such as the p in steady-state force balance. So far tokamak machines are the best design to demonstrate such equilibrium in plasma that we are looking for the purpose of MCF. However, the spherical torus is another option and yet the stellarator is another best option, and each can provide force balance for a reasonably high plasma pressure.

Stability in plasma can be depicted if Fig. 1.13 and in general a plasma equilibrium may be stable or unstable. Naturally from both words of expression, we can tell that stability is good and instability is bad in plasma confinement. However, effects of an MHD instability can be summarized as follows:

- Usually disastrous.
- Plasma moves and crashes into the wall.
- No more fusion.
- No more wall (in a reactor).
- This is known as a major disruption.

Fig. 1.13 Examples of stability

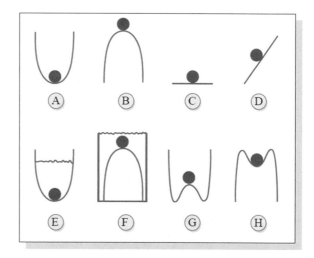

The job of MHD is to find magnetic geometries that stably confine high-pressure plasmas; large amount of theoretical and computational work has been done and well tested in experiments. The claim is that some say there is nothing left to do in fusion MHD based on the fact that the theory is essentially complete and computational tools are readily available and used routinely in experiments.

Although there is some truth in this view, however, still there are major unsolved MHD problems that need attention.

Historically, the MHD equations have been used extensively by astrophysicists working in cosmic electrodynamics, by hydrodynamicists working on MHD energy conversion, and by fusion scientist and theorists working with complicated magnetic geometries.

References

1. J.R. Reitz, F.J. Milford, R.W. Christy, *Foundations of Electromagnetic Theory*, 4th edn. (Pearson, Addison Wesley, Boston, MA, 2009)
2. F. Chen, *Introduction to Plasma Physics and Controlled Fusion*, 3rd edn. (Springer, New York, 2016)

Chapter 2
Principles of Plasma Physics

Physics of plasmas are a special class of gases made up of large number of electrons and ionized atoms and molecules, in addition to neutral atoms and molecules as are present in a non-ionized or so-called normal gas. Although by far most of the universe is ionized and is therefore in a plasma state, on our planet plasmas have to be generated by special processes and under special conditions. Since human got to know fire and thunderstorm caused lightning in the sky and the aurora borealis, we have been living in a bubble of essentially non-ionized gas in the midst of an otherwise ionized environment. The physics of plasma is a field in which knowledge is expanding rapidly, in particular a means of producing what is so known as source generating clean energy via either magnetic confinement or inertial confinement. The growing science of plasmas excites lively interest in many people with various levels of training.

2.1 Introduction

Given the complexity of plasma behavior, the field of plasma physics is best described as a web of overlapping models, each based on a set of assumptions and approximations that make a limited range of behavior analytically and computationally tractable.

A conceptual view of the hierarchy of plasma models/approaches to plasma behavior that will be covered in this text is shown in Fig. 2.1 here. We will begin with the determination of individual particle trajectories in the presence of electric and magnetic fields.

Subsequently, it will be shown that the large number of charged particles in plasma facilitates the use of statistical techniques such as plasma kinetic theory, where the plasma is described by a velocity-space distribution function. Quite often, the kinetic theory approach retains more information than we really want

© Springer International Publishing AG 2016
B. Zohuri, *Plasma Physics and Controlled Thermonuclear Reactions Driven Fusion Energy*, DOI 10.1007/978-3-319-47310-9_2

Fig. 2.1 Hierarchy of
approach to plasma
phenomena

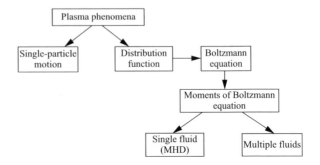

about a plasma, and a fluid approach is better suited, in which only macroscopic variables (e.g., density, temperature, and pressure) are kept.

The combination of fluid theory with Maxwell's equations forms the basis of the field of magnetohydrodynamics (MHD), which is often used to describe the bulk properties and collective behavior of plasmas. The remainder of this chapter reviews important physical concepts and introduces basic properties of plasmas.

We have all learned from our high school science that matter appears in three states, namely:

1. Solid
2. Liquid
3. Gaseous

However, in recent years, in particular after explosion of the thermonuclear weapon, scientists have paid more and more attention to the energy release control from such weapon as a new source of energy for our day-to-day use. Thus, their quest for new source of energy in a clean way (i.e., different than nuclear fission or coaled power plants) has taken them into different directions. This new direction has been toward controlled thermonuclear reactors, where deuterium (D) and tritium (T) fuse together to produce heavier nuclei such as helium and to liberate energy that can be found in our galaxy at the surface of the terrestrial universe. Therefore, for that reason they have looked into properties of matter at fourth and unique state, which is called *plasma*.

The higher the temperature, the more freedom the constituent particles of material experience.

In solid state of matter, the atoms and molecules are subject to strict solid and continuum mechanics discipline and are constructed to rigid order. In liquid form, matter can move, but their freedom is limited. However, at the stage of gaseous, they can move freely, and from the viewpoint of quantum mechanics laws, inside the atoms, the electrons perform a harmonic motion over their orbits.

However, matter in plasma stage is highly ionized and the electrons are liberated from atoms and acquire complete freedom of motion. Although plasma is often considered to be the fourth state of matter, it has many properties in common with the gaseous state. Meanwhile, the plasma is a fully ionized gas in which the long range of Coulomb forces gives rise to collective interaction effects, resembling a fluid with a density higher than that of a gas.

In its most general concept, plasma is any state of matter which contains sufficient free, charged particles for its dynamical behavior to be dominated by electromagnetic forces. Since atomic nuclei are positively charged, when two nuclei are brought together as a preliminary to combination or fusion, there is an increasing force of electrostatic repulsion of their positive charges, which is described by Coulomb and defined as Coulomb force and results in some barrier that is known as the Coulomb barrier.

In the fusion of light elements to form heavier ones, the nuclei (which carry positive electrical charge) must be forced close enough together to cause them to fuse into a single heavier nucleus. However, at a certain distance apart, the short-range nuclear attractive forces just exceed the long-range forces of repulsion, so the above fusion of the light elements becomes possible. The variation in the *potential energy* $V(r)$ of the system of two nuclei, with their distance r apart, is shown in Fig. 2.2 here.

Analyses of Fig. 2.2 indicate that a negative slope of potential energy curve presents net repulsion, whereas a positive slope implies net attraction. According to classical electromagnetic theory, the energy which must be supplied to the nuclei to surmount the Coulomb barrier, which is the amount of required energy to overcome the electrostatic repulsion so that fusion reaction can take place, is given by:

$$V(r) = \frac{Z_1 Z_2 e^2}{R_0} \tag{2.1}$$

where

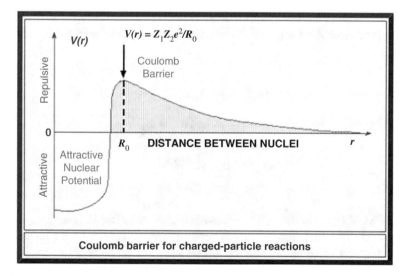

Fig. 2.2 Variation of the Coulomb potential energy with distance between nuclei

$V(r) = $ Potential energy to surmount Coulomb barrier

$Z_1 = $ Atomic number of nuclei element 1, carrying electric charge

$Z_2 = $ Atomic number of nuclei element 2, carrying electric charge

$e = $ Unit charge or Proton charge

$R_0 = $ Distance between the centers of element 1 and 2 at which the attractiveforces become dominant

As it was stated in the previous text, Fig. 2.2 indicates that the force between nuclei is repulsive until a very small distance separates them and then it rapidly becomes very attractive. Therefore, in order to surmount the Coulomb barrier and bring the nuclei close together where the strong attractive forces can be felt, the kinetic energy of the particles must be as high as the top of the Coulomb barrier.

In reality, effects associated with quantum mechanics help the situation. Because of what is termed as the Heisenberg uncertainty principle, even if the particles do not have enough energy to pass over the barrier, there is a very small probability that the particles will pass through the barrier. This is called barrier penetration or tunneling effect and is the means by which many such reactions take place in stars or terrestrial universe. Nevertheless, because this process happens with very small probability, the Coulomb barrier represents a strong hindrance to nuclear reactions in stars. Further discussion for barrier penetration can be found in the next section.

The key to initiating a fusion reaction is for the nuclei that are to fuse to collide at very high velocities, thus driving them close enough together for the strong (but very short-ranged) nuclear forces to overcome the electrical repulsion between them. In stars, the temperature and the density at the center of the star govern the probability of this happening.

For light nuclei, which are of interest for controlled thermonuclear fusion reactions, R_0 may be taken as approximately equal to a nuclear diameter, i.e., 5×10^{-13} em, and since e is 4.80×10^{-10} esu (statcoulomb), it follows from Equation 2.1 that:

$$
\begin{aligned}
V(r) &= \frac{Z_1 Z_2 e^2}{R_0} \\
&= \frac{\left(4.80 \times 10^{-10}\right)^2 Z_1 Z_2}{5 \times 10^{-13}} \\
&= 4.6 \times 10^{-7} Z_1 Z_2 \\
&= 0.28 Z_1 Z_2 \ \text{MeV}
\end{aligned}
\tag{2.2}
$$

where 1 MeV (million electron volts) is equivalent to 1.60×10^{-6} erg.

It is seen from Equation 2.2 that the energy within the nuclei must be acquired before they can be combined, increasing with the atomic numbers Z_1 and Z_2, and even for reactions among the isotopes of hydrogen, namely, deuterium (D) and tritium (H), for which $Z_1 = Z_2 = 1$, the minimum energy, according to classical theory, is about 0.28 MeV. Even larger energies should be reactions involving nuclei of higher atomic number because of the increased electrostatic repulsion. Although energies of the order of magnitude indicated by Equation 2.2 must be supplied to nuclei to cause them to combine fairly rapidly, experiments made with accelerated nuclei have shown that nuclear reactions can take place at detectable rates even when the energies are considerably below those corresponding to the top of the Coulomb barrier.

In other words, we cannot determine the threshold energy, by the maximum electrostatic repulsion of the interacting nuclei, below which the fusion reaction will not occur. Such behavior, which cannot be explained by way of classical mechanics, however, could be interpreted by means of wave mechanics [1].

2.2 Barrier Penetration

Classical physics reveals that a particle of energy E less than the height U_0 of barrier could not penetrate because the region inside the barrier is classically forbidden. However, the wave function associated with a free particle must be continuous at the barrier and will show an exponential decay inside the barrier. The wave function must also be continuous on the far side of the barrier, so there is a finite probability that the particle will tunnel through the barrier (see Fig. 2.3).

A free particle wave function in classical quantum mechanics is described as particle approaches the barrier. When it reaches the barrier, it must satisfy the Schrödinger equation in the form of a quantum harmonic oscillator as:

Fig. 2.3 Barrier penetration depiction

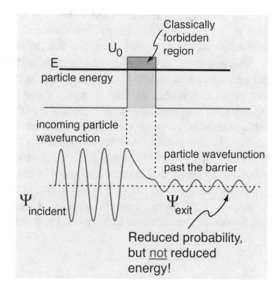

$$-\frac{\hbar^2}{2m}\frac{\partial^2 \Psi(x)}{\partial x^2} = (E - U_0)\Psi(x) \qquad (2.3a)$$

or

$$\frac{\partial^2 \Psi(x)}{\partial x^2} + \frac{2m(E - U_0)}{\hbar^2}\Psi(x) = 0 \qquad (2.3b)$$

Equation 2.3b is a one-dimensional ordinary differential equation that has the following solution as:

$$\Psi(x) = Ae^{-ax} \quad \text{where} \quad \alpha = \sqrt{\frac{2m(U_0 - E)}{\hbar^2}} \qquad (2.4)$$

where $\hbar = h/2\pi$ and h is Planck's constant.

Note that in addition to the mass and energy of the particle, there is a dependence on the fundamental physical constant Planck's constant h. Planck's constant appears in the Planck hypothesis where it scales the quantum energy of photons, and it appears in atomic energy levels, which are calculated using the Schrödinger equation.

2.3   Calculation of Coulomb Barrier

The height of the Coulomb barrier can be calculated if the nuclear separation and the charges of the particle are known. However, in order to accomplish nuclear fusion, the particles that are involved in this type of thermonuclear reaction must first overcome the electric repulsion Coulomb force to get close enough for the attractive nuclear strong force to take over to fuse with each other.

This requires extremely high temperatures, if temperature alone is considered in the process and one needs to calculate the temperature required to provide the given energy as an average thermal energy for each particle. Hence, a gas in thermal equilibrium has particles of all velocities, and the most probable distribution of these velocities obeys Maxwellian distribution, where we can calculate this thermal energy. In the case of the proton cycle in stars, this barrier is penetrated by tunneling, allowing the process to proceed at lower temperatures than that which would be required at pressures attainable in the laboratory.

Considering the barrier to be the electric potential energy of two point charges (e.g., point), the energy required to reach a separation r is given by the following relation as general form of Equation 2.1 and it is:

$$U = \frac{ke^2}{r} \qquad (2.5)$$

where k is Coulomb's constant and e is the proton charge. Given the radius r at which the nuclear attractive force becomes dominant, the temperature necessary to raise the average thermal energy to that point can be calculated.

The thermal energy is a physical notion of "temperature," which is *average translation kinetic energy* possessed by free particles given by *equipartition of energy*, which is sometimes called the thermal energy per particle. It is useful in making judgments about whether the internal energy possessed by a system of particles will be sufficient to cause other phenomena. It is also useful for comparisons of other types of energy possessed by a particle to that which it possesses simply as a result of its temperature. Additionally, from classical thermodynamics point of view, internal energy is the energy associated with the random, disordered motion of molecules. It is separated in scale from the macroscopic ordered energy associated with moving objects; it refers to the invisible microscopic energy on the atomic and molecular scale [2].

Note that the equipartition of energy theorem, states that molecules in thermal equilibrium have the same average energy associated with each independent degree of freedom of their motion and that the energy is:

$$\frac{1}{2}kT \text{ per molecule} \quad k = \text{Boltzmann's constant} \quad \frac{3}{2}kT$$

$$\frac{1}{2}RT \text{ per mole} \qquad R = \text{Gas constant} \qquad \frac{3}{2}RT$$

For three translational degrees of freedom, such as in an ideal mono-atomic gas, the above statements are true, and the equipartition result is then given by:

$$KE_{avg} = \frac{3}{2}kT \qquad (2.6)$$

Equation 2.6 serves well in the definition of kinetic temperature since that involves just the translational degrees of freedom, but it fails to predict the specific heats of polyatomic gases because the increase in internal energy associated with heating such gases adds energy to rotational and perhaps vibrational degrees of freedom. Each vibrational mode will get $kT/2$ for kinetic energy and $kT/2$ for potential energy—equality of kinetic and potential energy is addressed in the *virial theorem*. Equipartition of energy also has implication for electromagnetic radiation when it is in equilibrium with matter, each mode of radiation having kT of energy in the Rayleigh-Jeans law.

To prove the result of equipartition theory that is given by Equation 2.6 and follows the Boltzmann distribution, we do the following analyses. We easily derive this equation, by considering a gas in which the particles can move only in one dimension for the purpose of simplicity of the calculation, and in addition, we consider a strong magnetic field that can constrain electrons to move only the field lines; thus, the one-dimensional Maxwellian distribution is given by the following formula as:

$$f(u) = A\exp\left(-\frac{1}{2}mu^2/kT\right) \tag{2.7}$$

where $f(u)du$ is the number of particles per m^3 with velocity between u and $u + du$, where $mu^2/2$ is the kinetic energy and k is Boltzmann constant and its value is equal to 1.38×10^{-23} J/K. The constant A is related to particle density A, as it is shown below:

$$A = n\left(\frac{m}{2\pi kT}\right)^{1/2} \tag{2.8}$$

and this density is analyzed below.

Using Fig. 2.4, we can write the formula for particle density n, or the number of particles per m^3 is given by:

$$n = \int_{-\infty}^{+\infty} f(u)du \tag{2.9}$$

The width of the distribution in Fig. 2.4 is characterized by the constant T, which we call the temperature. To have a concept of the exact meaning of temperature T, we can compute the average kinetic energy of particles within this distribution:

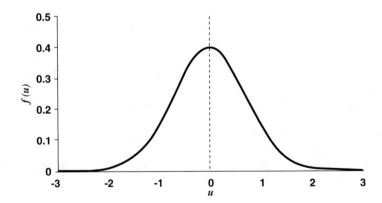

Fig. 2.4 A Maxwellian velocity distribution

$$KE_{avg} = \frac{\int_{-\infty}^{+\infty} \frac{1}{2} mu^2 f(u) du}{\int_{-\infty}^{+\infty} f(u) du} \qquad (2.10)$$

Defining a new variable v_{th} as below:

$$v_{th} = (2kT/m)^{1/2} \qquad (2.11)$$

Substitution of this new variable results in Equations 2.7 and 2.10 to become as:

$$f(u) = A\exp\left(\frac{u^2}{v_{th}}\right)$$

$$KE_{avg} = \frac{\frac{1}{2} mA v_{th}^3 \int_{-\infty}^{+\infty} [\exp(-y^2)] y^2 dy}{A v_{th} \int_{-\infty}^{+\infty} [\exp(-y^2)] dy} \qquad (2.12)$$

The integral in the numerator, in the second equation set of Equation 2.12, is integrable by parts as:

$$\int_{-\infty}^{+\infty} y \cdot [\exp(-y^2)] y dy = \{-\tfrac{1}{2}[\exp(-y^2)] y\}_{-\infty}^{+\infty} - \int_{-\infty}^{+\infty} \frac{1}{2} \exp(-y^2) dy$$
$$= \frac{1}{2} \int_{-\infty}^{+\infty} xp(-y^2) dy = A v_{th} \qquad (2.13)$$

Substituting Equation 2.13 into Equation 2.12 and canceling the common integrals from denominator and numerator result in the following relation for average kinetic energy as:

$$KE_{avg} = \frac{1}{2} mA v_{th}^3 \cdot \frac{\frac{1}{2}}{A v_{th}} = \frac{1}{4} m v_{th}^2 = \frac{1}{2} kT \qquad (2.14)$$

which is proof of equipartition scenario and indicates that the average kinetic energy is $\frac{1}{2} kT$. By similar analyses, Equation 2.14 can be easily expanded to three-dimensional form; therefore, the Maxwellian distribution for three-dimensional Cartesian coordinate system becomes:

$$f(u, v, w) = A_3 \exp\left[\frac{1}{2}(u^2 + v^2 + w^2)/kT\right] \qquad (2.15)$$

where constant A_3 is given as:

$$A_3 = n\left(\frac{m}{2\pi kT}\right)^{3/2} \tag{2.16}$$

In that case, the average kinetic energy KE_{avg} is presented by:

$$KE_{avg} = \frac{\displaystyle\iint\int_{-\infty}^{+\infty}(A_3)\frac{1}{2}m(u^2 + v^2 + w^2)\exp\left[-\frac{1}{2}(u^2 + v^2 + w^2)/kT\right]dudvdw}{\displaystyle\iint\int_{-\infty}^{+\infty}A_3\exp\left[-\frac{1}{2}(u^2 + v^2 + w^2)/kT\right]dudvdw} \tag{2.17}$$

Equation 2.17 is symmetric in variables u, v, and w, since a Maxwellian distribution is isotropic. As a result, each of the three terms in the numerator is the same as the others; hence, all we need to do is to evaluate the first and multiply it by three and get the following result as:

$$KE_{ave} = 3A_3\frac{\displaystyle\int\frac{1}{2}mu^2\exp\left(-\frac{1}{2}mu^2/kT\right)du\iint\exp\left[-\frac{1}{2}m(v^2 + w^2)/kT\right]dvdw}{\displaystyle A_3\int\exp\left(-\frac{1}{2}mu^2/kT\right)du\iint\exp\left[-\frac{1}{2}m(v^2 + w^2)/kT\right]dvdw} \tag{2.18}$$

Using our previous result, we have:

$$KE_{ave} = \frac{3}{2}kT \tag{2.19}$$

Solution to Equation 2.19 and mathematical process for obtaining it prove Equation 2.6, and it is an indication that KE_{ave} in general equals to $\frac{1}{2}kT$ per degree of freedom.

Since temperature T and average kinetic energy KE_{ave} are so closely related, it is customary in plasma physics to give temperatures in units of energy. To avoid confusion on the number of dimensions involved, it is not KE_{ave} but the energy corresponding to kT that is used to denote the temperature. For $kT = 1\,eV = 1.6 \times 10^{-19}\,J$, we have:

$$T = \frac{1.6 \times 10^{-19}}{1.38 \times 10^{-23}} = 11,600 \tag{2.20}$$

Thus, the conversion factor is:

$$1\,eV = 11,600\,K$$

2.4 Thermonuclear Fusion Reactions

As part of the thermonuclear fusion reaction system, we have to have some understanding of energies related to the reacting nuclei that are following a Maxwellian distribution and the problem in hand; this distribution can be presented by:

$$dn = \text{constant} \times \frac{E^{1/2}}{T^{13/2}} \exp\left(-\frac{E}{kT}\right) dE \tag{2.21}$$

where dn is the number of nuclei per unit volume whose energies, in the frame of the system, lie in the range from E to $E + dE$, k is again Boltzmann constant and is equal to 1.38×10^{-16} erg/K, and T is the *kinetic temperature*. The kinetic temperature of a system of particles is defined as the temperature appropriate to Maxwellian distribution assumed by the particles upon equipartition of energy with three translational degrees of freedom. The mean particle energy is then $3kT/2$ as it was defined by Equation 2.19.

Incidentally, when a system is in blackbody radiation equilibrium, per description given by Glasstone and Lovberg [1], the radiation pressure is equal to aT^4/c, where c is the velocity of light. For a temperature of 10 KeV, i.e., 1.16×10^8 K, this would be the order of 10^{11} atm. In stars, such high pressures are balanced by gravitational forces due to the enormous masses. Naturally, there exists no practical controlled thermonuclear reactor that could withstand the pressure resulting from equilibrium with radiation at extremely high temperature. Thus, the solution around this problem is by utilization of the very low particle densities required by other considerations. A system of this type is optically "thin" and transparent to essentially all the Bremsstrahlung emission from a hot plasma; it is a poor absorber and hence also a poor emitter of this radiation. The radiation field with which the particles may be in equilibrium is then very much weaker than blackbody radiation. In other words, the equivalent radiation temperature is much lower than kinetic temperature, which is related to the energy distribution among the particles [1].

It is for this reason that the term kinetic temperature, rather than just temperature without qualification, has been frequently used in the preceding text. Strictly speaking, "temperature" implies thermodynamic equilibrium, which means both kinetic and blackbody radiation equilibrium [2].

Theoretically it has been proven that the energy of the interacting particles represented by the atomic numbers of Z_1 and Z_2, with individual mass m_1 and m_2, is well below the top of the Coulomb barrier. In addition, the cross section for the combination of two nuclei can be written down to a good approximation in the form of Equation 2.22, as a function of the relative particle energy E, which represents the total kinetic energy of the two nuclei in the center-of-mass system as:

$$\sigma(E) \approx \frac{\text{Constant}}{E^{3/2}} \cdot \exp\left[-\frac{2^{3/2}\pi^2 M^{1/2}Z_1 Z_2 e^2}{hE^{1/2}}\right] \tag{2.22}$$

where h is the Planck's constant and M is the reduced mass of two individual particles interacting with each other and expressed as:

$$M = \frac{m_1 m_2}{m_1 + m_2} \tag{2.23}$$

Moreover, Equation 2.22 reveals that the fusion reaction has a finite cross section, even when the relative energy of the nuclei is quite small; however, because the exponential term in that equation is a dominating factor, the cross section increases rapidly as the relative particle energy increases. It can be noted as well that for a given value of the relative energy, the reaction cross section decreases with increasing atomic number of the interacting nuclei [1].

The contribution to the overall reaction rate per unit energy interval made by nuclei with relative energy in the range from E to $E + dE$ in a thermonuclear system, at the kinetic temperature T, is proportional to the product of $\sigma(E)$ and of dn/dE for that particular temperature. This contribution may be expressed by $R(E)$, so that from Equations 2.21 and 2.22:

$$R(E) \approx \frac{C}{E^{3/2}T^{3/2}}\exp\left[-\frac{2^{3/2}\pi^2 M^{1/2}Z_1 Z_2 e^2}{hE^{1/2}} - \frac{E}{kT}\right] \tag{2.24}$$

where C is a constant [1].

Figure 2.5 shows the significance of Equation 2.24 here, and the curve marked dn/dE is a typical Maxwellian distribution of the relative particle energies for a specified kinetic temperature.

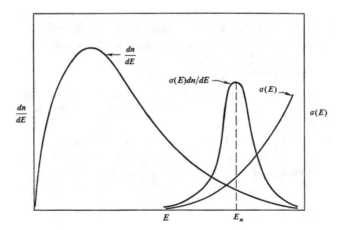

Fig. 2.5 Effect of Maxwellian energy distribution on nuclear reaction rate

The cross section variation for the nuclear fusion reaction with the relative energy, as determined by Equation 2.22, is shown by curve $\sigma(E)$. The dependence of the contribution to the reaction rate made by particles of relative energy E, obtained by multiplying the ordinates of the other two curves, is indicated by the curve in the center of Fig. 2.5. It can be seen that this curve has a distinct maximum corresponding to the relative energy E_m, so that nuclei having this amount of relative energy make the maximum contribution to the total fusion reaction rate [1].

The average energy of the nuclei is in the vicinity of the maximum of the dn/dE curve; therefore, it is evident that E_m is larger than the average energy for the given kinetic temperature.

Hence, in order to determine the total reaction, we need to determine the total area under the curve of $\sigma(E)dn/dE$ in Fig. 2.5 by integrating over the function curve $\sigma(E)dn/dE$, from energy point 1 to point 2. Consequently, it is obvious that most of the considered thermonuclear reaction will be due to a relatively small fraction of the nuclear collisions in which the relative energies are greatly in excess of the average.

The preceding text explains why there is an advantage in performing a nuclear fusion reaction, e.g., with uniformly accelerated particles, to permit the nuclei to become "thermalized," that is, to attain a Maxwellian distribution of energies, as a result of collision.

Now, we can calculate the maximum relative energy E_m by differentiating Equation 2.24 with respect to energy E, providing that the kinetic temperature is not too high. However, the variation $R(E)$ with E is determined almost entirely by the exponential factor in Equation 2.24; hence, a good approximation to the value of E_m in Fig. 2.5 can be obtained by calculating the energy for which this factor is a maximum, and this value is found to be:

$$E_m \approx \left[\frac{(2M)^{1/2}\pi^2 Z_1 Z_2 e^2 kT}{h} \right]^{2/3} \qquad (2.25)$$

The expression in Equation 2.25 gives the relative energy in nuclear collision making the maximum contribution to the reaction rate at the not too high kinetic temperature T. Dividing both sides of Equation 2.25 by kT, we obtain the following result as:

$$\frac{E_m}{kT} \approx \left[\frac{(2M)^{1/2}\pi^2 Z_1 Z_2 e^2}{h} \right]^{2/3} \frac{1}{(kT)^{1/3}} \qquad (2.26)$$

Note that it is a common practice in thermonuclear studies to express kinetic temperature in terms of the corresponding energy kT in kilo-electron volts, i.e., in KeV units. Since the Boltzmann constant k is 1.38×10^{-16} erg/K and 1 keV is equivalent to 1.60×10^{-9} erg, it follows that [1]:

$$k = 8.6 \times 10^{-8} \text{ KeV/K}$$

or

$$1 \text{ keV} = 1.16 \times 10^7 \text{ K}$$

Thus a temperature of T keV is equivalent to $1.16 \times 10^7 T$ K.

2.5 Rates of Thermonuclear Reactions

Among all the related text to this particular subject that I have personally seen, the best book that describes the rates of thermonuclear reactions is given by Glasstone and Lovberg [1]; consequently, I will use exactly what they have described for this matter.

Consider a binary reaction in a system containing n_1 nuclei/cm^3 of one reacting element and n_2 of the other. To determine the rate at which the two nuclear elements interact, it may be supposed that the nuclei of the first element kind form a stationary lattice within the nuclei of the second kind which move at random with a constant velocity v cm/s equal to the relative velocity of the nuclei. The total cross section for all the stationary nuclei in 1 cm^3 is then $n_1\sigma$ nuclei/cm. This gives the number of nuclei of the first kind with which each nucleus of the second kind will react while traveling a distance of 1 cm. The total distance traversed in 1 s by all the nuclei of the latter type present in 1 cm^3 is equal to n_2v nuclei/(cm^2)(s). Hence, the nuclear reaction rate R_{12} is equal to the product of $n_1\sigma$ and n_2v; thus,

$$R_{12} = n_1 n_2 \sigma v \quad \text{Interaction/(cm}^2)(s) \tag{2.27}$$

If the reaction occurs between two nuclei of the same kind, e.g., two Deuterons, so that n_1 (n sub 1) and n_2 (n sub 2) are equal, the expression for the nuclear reaction rate, represented by R_{11}, becomes:

$$R_{11} = \frac{1}{2}n^2 \sigma v \quad \text{Interaction/(cm}^2)(s) \tag{2.28}$$

where n is the number of reactant nuclei/cm^3 (see Fig. 2.6).

In order that each interaction between identical nuclei is not going to be counted twice, the factor of 1/2 is introduced into Equation 2.28.

Going forward, the two established Equations 2.27 and 2.28 are applicable when the relative velocity of the interacting nuclei is constant, as is true, approximately at least, for high-energy particle from an accelerator. However, for thermonuclear reaction, there would be a distribution of velocities and energies as well, over a wide range.

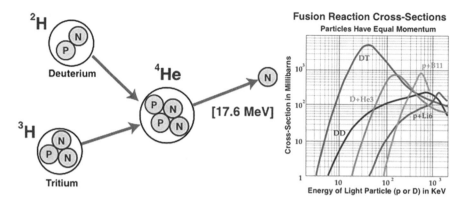

Fig. 2.6 Depiction of all isotopes of hydrogen thermonuclear reactions

As it is depicted in Fig. 2.6 on the right-hand side, it shows that the reaction cross section is dependent on the energy or velocity, and generally speaking it follows the product σv in Equations 2.27 and 2.28 that needs to be replaced by a value such as symbol of $\overline{\sigma v}$, which is averaged over the whole range of relative velocities. Thus, Equation 2.27 is written as:

$$R_{12} = n_1 n_2 \overline{\sigma v} \quad \text{Interaction}/(\text{cm}^2)(\text{s}) \tag{2.29}$$

Accordingly, Equation 2.28 becomes:

$$R_{11} = \frac{1}{2} n^2 \overline{\sigma v} \tag{2.30}$$

Using reduced mass M expressed by Equation 2.23, which is the result of the interaction between two individual masses, two elements can be used to describe the new form of Equation 2.21, providing that the velocity distribution is Maxwellian and we know that the kinetic energy is $E = Mv^2/2$. Thus, we can write:

$$dn = n \left(\frac{M}{2\pi kT} \right)^{3/2} \exp\left(-\frac{Mv^2}{2kT} \right) v^2 dv \tag{2.31}$$

where dn is the number of particles whose velocities relative to that of a given particle lie in the range from v to $v + dv$. Hence, it follows that:

$$\overline{\sigma v} = \frac{\int_0^\infty \sigma v \, dn}{\int_0^\infty dn} = \frac{\int_0^\infty \sigma v \left[\exp\left(-\frac{Mv^2}{2kT} \right) v^2 dv \right]}{\int_0^\infty \exp\left(-\frac{Mv^2}{2kT} \right) v^2 dv} \tag{2.32}$$

The integral in the denominator of Equation 2.32 is equal to $[(2kT/M)^{3/2}](\pi^{1/2}/4)$, and so this equation becomes:

$$\overline{\sigma v} = \frac{4}{\pi^{1/2}} \left(\frac{Mv^2}{2kT}\right) \int_0^\infty \sigma \exp\left(-\frac{Mv^2}{2kT}\right) v^2 \, dv \tag{2.33}$$

The integral term in Equation 2.33 can be evaluated by changing the variable. Since nuclear cross sections are always determined and expressed as a function of the energy of the bombarding particle, the bombarded particle being essentially at rest in the target, the actual velocity of the bombarding nucleus is also its relative velocity. Hence, if E is the actual energy, in the laboratory system, of the bombarding nucleus of mass m, then E is written as:

$$E = \frac{1}{2} m v^2 \tag{2.34a}$$

So that

$$v = \left(\frac{2E}{m}\right)^{1/2} \tag{2.34b}$$

And differentiating both sides of Equation 2.34b, we get:

$$v^2 \, dv = \frac{2E}{m^2} \, dE \tag{2.34c}$$

Substitution of Equation 2.34c into Equation 2.33 yields:

$$\overline{\sigma v} = \frac{4}{\pi^{1/2}} \left(\frac{M}{2kT}\right)^{3/2} \frac{1}{m^2} \int_0^\infty \sigma \exp\left(-\frac{ME}{mkT}\right) E \, dE \tag{2.35}$$

where σ in the integrand is the cross section for a bombarding nucleus of mass m and energy E.

If the temperature T in Equation 2.35 is expressed in kilo-electron volts and the values of E are in the same units, it is convenient to rewrite Equation 2.35 in the new form as:

$$\overline{\sigma v} = \left(\frac{8}{\pi^{1/2}}\right)^{1/2} \frac{M^{3/2}}{m^2} \int_0^\infty \sigma \exp\left(-\frac{ME}{mT}\right) \frac{E}{T} \, dE \tag{2.36}$$

where the quantity E/T is dimensionless. If σ, determined experimentally, can be expressed as a relatively simple function of E, as is sometimes the case, the integration in Equation 2.36 may be performed analytically. Alternatively, numerical methods, for example, Simpson's rule, may be employed.

In any event, the values of $\overline{\sigma v}$ for various kinetic temperatures can be derived from Equation 2.36, based on Maxwellian distribution of energies or velocities, and the results can be inserted in Equation 2.29 or 2.30 to give the rate of a thermonuclear reaction at a specified temperature.

2.6 Thermonuclear Fusion Reactions

In a thermonuclear fusion reactions, two light nuclei are forced together, and then they will fuse with a yield of energy as it is depicted in Fig. 2.7. The reason behind the energy yield is due to the fact that the mass of the combination of fusion reaction will be less than the sum of the masses of the individual nuclei.

If the combined nuclear mass is less than that of iron at the peak of the binding energy curve, then the nuclear particles will be more tightly bound than they were in the lighter nuclei and that decrease in mass comes off in the form of energy according to the Einstein relationship. However, elements heavier than iron fission reaction will yield energy.

The Einstein relationship that is known as theory of relativity indicates that relativistic energy is presented as:

$$E = mc^2 \tag{2.37}$$

where m is the effective relativistic mass of particle traveling at a very high of speed c. Equation including both the kinetic energy and rest mass energy m_0 for a particle can be calculated from the following relation:

$$KE = mc^2 - m_0c^2 \tag{2.38}$$

Further analysis of Einstein relativity theory allows us to blend into Equation 2.38, the relativistic momentum p expression as:

Fig. 2.7 A thermonuclear fusion reaction

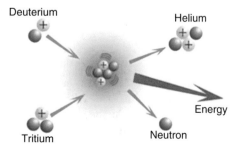

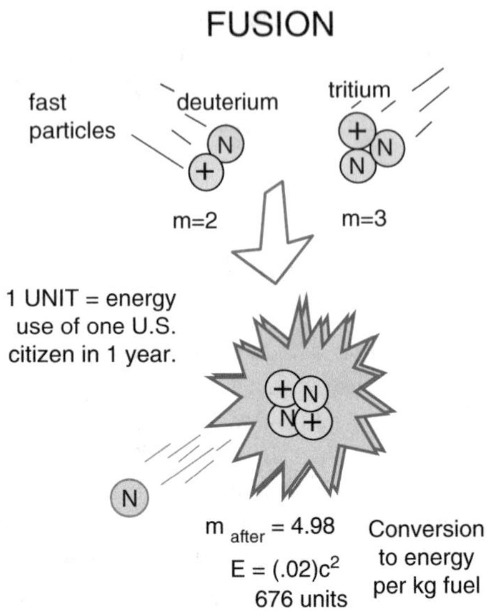

Fig. 2.8 Deuterium-tritium
fusion reaction

FUSION

fast
particles

deuterium

tritium

m=2   m=3

1 UNIT = energy
use of one U.S.
citizen in 1 year.

$m_{after} = 4.98$ Conversion
$E = (.02)c^2$ to energy
676 units   per kg fuel

$$p = \frac{m_0 v}{\sqrt{1 - \frac{v^2}{c^2}}} \qquad (2.39)$$

The combination of relativistic momentum p and particle speed c shows up often in relativistic quantum mechanics and relativistic mechanics as multiplication of pc, and it can be manipulated as follows, using conceptual illustration such as Fig. 2.8:

$$p^2 c^2 = \frac{m_0^2 v^2 c^2}{1 - \frac{v^2}{c^2}} = \frac{m_0^2 \frac{v^2}{c^2} c^4}{1 - \frac{v^2}{c^2}} \qquad (2.40a)$$

and by adding and subtracting a term, it can be put in the form:

$$p^2 c^2 = \frac{m_0^2 c^4 \left[\frac{v^2}{c^2} - 1\right]}{1 - \frac{v^2}{c^2}} + \frac{m_0^2 c^4}{1 - \frac{v^2}{c^2}} = -m_0^2 c^4 + \left(mc^2\right)^2 \qquad (2.40b)$$

which may be rearranged to give the following expression for energy:

$$E = \sqrt{p^2 c^2 + \left(m_0 c^2\right)^2} \qquad (2.40c)$$

Note that again m_0 is the rest mass and m is the effective relativistic mass of particle of interest at very high speed c.

Per Equation 2.40c, the relativistic energy of a particle can also be expressed in terms of its momentum in the expression such as:

$$E = mc^2 = \sqrt{p^2c^2 + m_0^2c^4} \tag{2.41}$$

The relativistic energy expression is the tool used to calculate binding energies of nuclei and energy yields of both nuclear fission and thermonuclear fusion reactions.

Bear in your mind that the nuclear binding energy is rising from the fact that nuclei are made up of proton and neutron, but the mass of a nucleus is always less than the sum of the individual masses of the protons and neutrons, which constitute it. The difference is a measure of the nuclear binding energy, which holds the nucleus together. This binding energy can be calculated from the Einstein relationship:

$$\text{Nuclear Binding Energy} = \Delta mv^2 \tag{2.42}$$

Now that we have a better understanding of the physics of thermonuclear fusion reaction and we explained what the Coulomb barriers and energy are all about, we pay our attention to thermonuclear fusion reaction of hydrogen, which is the fundamental chemical element of generating energy-driven controlled fusion.

According to Glasstone and Lovberg [1]:

> "because of the increased height of the Coulomb energy barrier with increasing atomic number, it is generally true that, at a given temperature, reactions involving the nuclei of hydrogen isotopes take place more readily than do analogous reactions with heavier nuclei. In view of the greater abundance of the lightest isotope of the hydrogen, with mass number 1, it is natural to see if thermonuclear fusion reactions involving this isotope could be used for the release of energy" [1].

Unfortunately, the three possible reactions between hydrogen (H) nuclei alone and with deuterium (D) or tritium (T) nuclei, i.e.:

$$_1H^1 + {}_1H^1 \rightarrow {}_1D^2 + {}_1e^0$$
$$_1H^1 + {}_1D^2 \rightarrow {}_1D^2 + \gamma$$
$$_1H^1 + {}_1T^3 \rightarrow {}_2He^4 + \gamma$$

are known to have cross sections that are too small to permit a net gain of energy at temperature which may be regarded as attainable [1].

Consequently, recourse must be the next most abundant isotope, i.e. deuterium, and here two reactions, which occur at approximately the same rate over a considerable range of energies, are of interest; these are the D-D reactions as:

$$_1D^2 + {}_1D^2 \rightarrow {}_2He^3 + {}_1n^0 + 3.27\ \text{MeV}$$

and

$$_1D^2 + {}_1D^2 \rightarrow {}_2T^3 + {}_1H^0 + 4.03\ \text{MeV}$$

called the "neutron branch" and the "proton branch," respectively. The tritium produced in the proton branch or obtained in another way, as explained below, can then react, at a considerably faster rate, with deuterium nuclei in the D-T reaction as:

$$_1D^2 + \, _1T^3 \rightarrow \, _2He^4 + \, _1n^0 + 17.60 \, \text{MeV}$$

The He^3 formed in the first D-D reaction can also react with deuterium; thus:

$$_1D^2 + \, _1He^3 \rightarrow \, _2He^4 + \, _1H^0 + 18.30 \, \text{MeV}$$

This reaction is of interest because, as in the D-T reaction, there is a large energy release; the D-He^3 reaction is, however, slower than the other at low thermonuclear temperatures, but its rate approached that of the D-D reactions at 100 KeV and demonstrated in Fig. 2.9.

In the methods currently under consideration for production of thermonuclear power, the fast neutrons produced in neutron branch of the D-D reactions and in the D-T reactions would most probably escape from the immediate reaction environment. Thus, a suitable moderator that can be considered to slow down these neutrons can be either water, lithium, or beryllium; with the liberation of their kinetic energy in the form of heat, they can be utilized:

$$_3Li^6 + \, _0n^1 \rightarrow \, _2He^4 + \, _1T^3 + 4.6 \, \text{MeV}$$

The slow neutrons can then be captured in lithium-6, which constitutes 7.5 at.% of natural lithium, by the reaction in the above, leading to the production of tritium. The energy released can be used as heat, and the tritium can, in principle, be transferred to the thermonuclear system to react with deuterium.

If we produce enough initial ignition temperature to the above four thermonuclear reactions, all four fusion processes will take place, and the two neutrons produced would subsequently be captured by lithium-6.

By means of the quantum mechanics theory of the Coulomb barrier penetration, it is much more convenient to make use of cross sections obtained experimentally as it is plotted in Fig. 2.9 for reactions such as D-D, D-T, and D-He^3, by bombarding targets containing deuterium, tritium, or helium-3 with deuterons of known energies. Technically, for the purpose of marginal safety measurements of the cross section, it is normally done with order-of-magnitude estimation, at least, of the rates or cross section of thermonuclear reactions obtained experimentally.

It will be observed that the D-T curve demonstrates a maximum energy of 110 keV, which is an example of the resonance phenomenon, which often occurs in nuclear reactions [1].

However, the appreciable cross sections for energies well below the top of the Coulomb barriers for each of the reaction studies provide an experimental illustration of the reality of the barrier penetration effect.

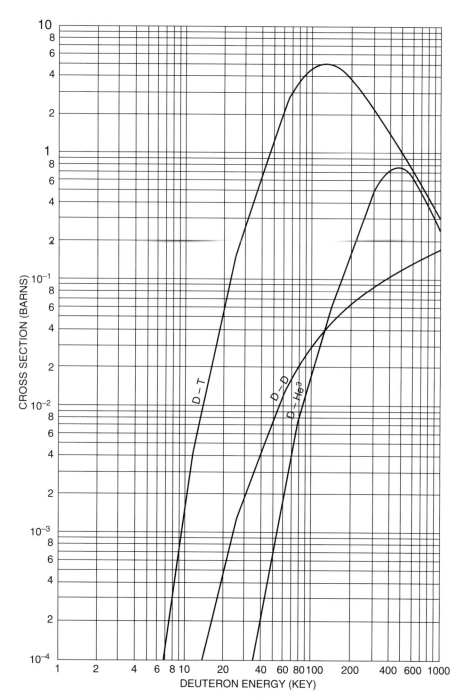

Fig. 2.9 Cross sections for D-T, D-D total, and D-He^3 reactions

The data in Fig. 2.9, for particular deuteron energies, are applied to the determination of the average $\overline{\sigma v}$ that is presented by Equations 2.35 and 2.36, assuming a Maxwellian distribution of particle energies or velocities. Figure 2.10 here shows the result of integration presented by Equation 2.36 and the curve that gives $\overline{\sigma v}$ in cm^3/s as a function of the kinetic temperature of the reaction system in kilo-electron volts. The values in the plot in Fig. 2.10 for a number of temperatures are also marked in Table 2.1 here.

Figures 2.9 and 2.10 both illustrate the overall effect on the thermonuclear fusion reaction rates that are taking into account the Maxwellian distribution.

Analytical expression for σ and $\overline{\sigma v}$ for the D-D and D-T fusion reactions can be obtained by utilizing Equation 2.24 in a somewhat modified form. The relative kinetic energy E of the nuclei is given as:

$$E = \frac{1}{2}Mv^2 \tag{2.43}$$

where v is the relative velocity, and the deuteron energy E_D, in terms of which the cross section are expressed, is $m_D v^2/2$, where m_D is the mass of the deuteron. Hence, $(M/E)^{1/2}$ in Equation 2.24 may be replaced by $(m_D/E_D)^{1/2}$; since Z_1 and Z_2 are both unity, the result then is:

$$\begin{aligned}
\sigma(E_D) &= \frac{C}{E_D}\exp\left[-\frac{2^{3/2}\pi^2 m_D^{1/2}e^2}{hE_D^{1/2}}\right] \\
&= \frac{C}{E_D}\exp\left[-\frac{44.24}{E_D^{1/2}}\right]
\end{aligned} \tag{2.44}$$

with E_D expressed in kilo-electron volts. Note that the potential factor is the same for both D-D and D-T thermonuclear fusion reactions, with the deuteron as the projectile particle. The factor preceding the exponential will, however, be different in the two cases [1].

Now if we are interested in mean free path reaction λ, in a system containing n nuclei/cm^3 of a particular reacting species, then λ is the average distance traveled by a nucleus before it undergoes reaction and is equal to $1/n\sigma$, where σ is the cross section for the given reaction [1].

We replace σ with $\overline{\sigma}$, if a Maxwellian distribution is considered, and in this case the averaged cross section $\overline{\sigma}$ is taken over all energies from zero to infinity, at a given kinetic temperature.

Figure 2.11 is an illustration of the mean free path values for a deuteron in centimeter as a function of the deuteron particle density n, in nuclei/cm^3, for the D-D and D-T reactions at two kinetic temperatures, 10 and 100 KeV, in each case, and temperatures of these orders of magnitude would be required in a controlled thermonuclear fusion reactor.

The particle of interest for possible fusion reaction for controlled thermonuclear process has a possible density of about 10^{15} deuterons/cm^3; the mean free path at

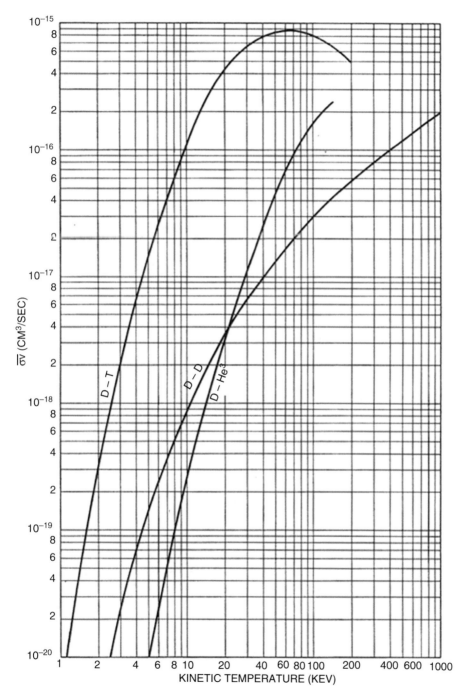

Fig. 2.10 Values of $\overline{\sigma v}$ based on Maxwellian distribution for D-T, D-D (total), and D-He^3 reactions

Table 2.1 Values of $\overline{\sigma v}$ at specified kinetic temperature

Temperature (KeV)	D-D (cm³/s)	D-T (cm³/s)	D-He³ (cm³/s)
1.0	2×10^{-22}	7×10^{-21}	6×10^{-28}
2.0	5×10^{-21}	3×10^{-19}	2×10^{-23}
5.0	1.5×10^{-19}	1.4×10^{-17}	1×10^{-20}
10.0	8.6×10^{-19}	1.1×10^{-16}	2.4×10^{-19}
20.0	3.6×10^{-18}	4.3×10^{-16}	3.2×10^{-19}
60.0	1.6×10^{-17}	8.7×10^{-16}	7×10^{-17}
100.0	3.0×10^{-17}	8.1×10^{-16}	1.7×10^{-18}

100 KeV for the D-D reaction, according to Fig. 2.11, is about 2×10^{16} cm. This statement translates to the fact that, at the specified temperature and particle density, a deuteron would travel on the average distance of 120,000 miles before reacting. For D-T reaction the mean free paths are shorter, because of the large cross sections for deuterons of given energies, but they are still large in comparison with the dimensions of normal equipment. All these results play a great deal of impotency to the problem of confinement of the particles in a thermonuclear fusion reacting system such as tokamak machine or any other means.

For the purpose of obtaining a power density P_{DD} of thermonuclear fusion reaction, such as D-D, we use either Equation 2.29 or 2.30 to calculate the rate of thermonuclear energy production. If we assume an amount of average energy Q in erg is produced per nuclear interaction, then using Equation 2.20, it follows that:

$$\text{Rate of energy release} \; = \; \frac{1}{2}n_D^2\overline{\sigma v}Q\,\text{ergs}/\left(\text{cm}^3\right)(\text{s}) \tag{2.45}$$

If the dimension of power density P_{DD} is given in W/cm³, which is equal to 10^7 ergs/(cm³)(s), then we can write:

$$P_{DD} = \frac{1}{2}n_D^2\overline{\sigma v}Q \times 10^{-7} \tag{2.46}$$

with n_D in deuterons/cm³, $\overline{\sigma v}$ in units of cm³/s, and average energy Q in erg.

For every two D-D interactions, an average of 8.3 MeV of energy is deposited within the reacting system. The energy Q per interaction is thus, $(1/2) \times 8.3 \times 1.60 \times 10^{-6} = 6.6 \times 10^{-6}$ erg, and upon substitution into Equation 2.46, it yields that:

$$P_{DD} = 3.3 \times 10^{-13}n_D^2\overline{\sigma v}\,\text{W}/\text{cm}^3 \tag{2.47}$$

As an example for utilization of 2.46, we look at a D-D reaction at 10 KeV, and from Fig. 2.10 or Table 2.1, for a given kinetic temperature, we see that $\overline{\sigma v}$ is equal to 8.6×10^{-19} cm³/s; therefore, the power density is:

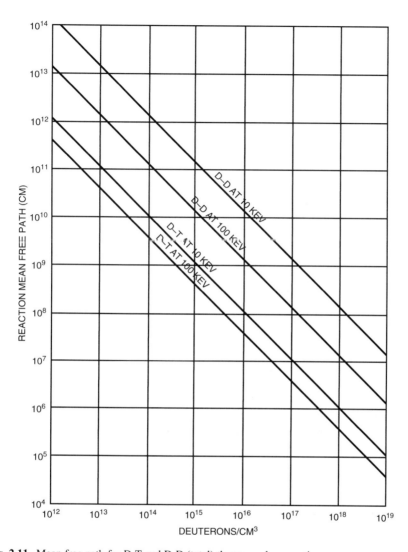

Fig. 2.11 Mean free path for D-T and D-D (total) thermonuclear reactions

$$P_{DD}(10\,\text{KeV}) = 2.8 \times 10^{-31} n_D^2 \, \text{W/cm}^3 \tag{2.48}$$

and at 100 KeV, when $\overline{\sigma v}$ is equal 3.0×10^{-17} cm^3/s, the power density will be:

$$P_{DD}(100\,\text{KeV}) = 10^{-29} n_D^2 \, \text{W/cm}^3 \tag{2.49}$$

Similar analysis can be performed for thermonuclear reaction fusion reaction of D-T, knowing that the energy remaining in the system per interaction is 3.5 MeV, i.e., $3.5 \times 1.6 \times 10^{-6}$ erg, then the reaction rate is given by Equation 2.29, and therefore, the thermonuclear reactor density power is:

$$P_{DT} = \frac{1}{2} n_D n_T \overline{\sigma v} \, Q \times 10^{-7} \qquad (2.50)$$

where in this case, the average energy Q is 5.6×10^{-6} erg; hence,

$$P_{DT}(10\,\text{KeV}) = 6.2 \times 10^{-13} n_D n_T \, \text{W/cm}^3 \qquad (2.51)$$

and

$$P_{DT}(100\,\text{KeV}) = 4.5 \times 10^{-28} n_D n_T \, \text{W/cm}^3 \qquad (2.52)$$

There is no exact parallel correlation between the conditions of heat transfer and operating pressures, which limit the power density of a fission reactor and those, which might apply to a thermonuclear fusion reactor. Nevertheless, there must be similar limitations upon power transfer in a continuously operating thermonuclear reactor as in other electrical power systems.

A large steam-powered electrical generating plant has a power of about 500 MW, i.e., 5×10^8 W. Figure 2.12 is illustrating that a 100 KeV in a D-D reactor that has a power of 5×10^8 W would provide a reacting volume of only 0.03 cm^3 with deuteron particle densities equivalent to those at standard temperature.

Meanwhile, the gas kinetic pressure exerted by the thermonuclear fuel would be about 10^7 atm or 1.5×10^8 psi. Since the mean reaction lifetime is only a few milliseconds under the conditions specified, it is obvious that the situation would be completely impractical [1].

From what have discussed so far, it seems that the particle density in a practical thermonuclear reactor must be near 10^{15} nuclei/cm^3. Other problems are associated with the controlled thermonuclear fusion reaction for plasma confinement and that is why the density cannot be much larger, and it can be explained via stability requirements that are frequently restricted by dimensionless ratio β. This ratio is defined as part of convenience in plasma confinement driven by magnetic field, which is equal to the kinetic pressure of the particles in plasma in terms of its ratio to the external magnetic pressure or energy density, which is defined by Equation 3.85 of this book.

Details of this dimensionless parameter will be defined toward the end of this chapter as well.

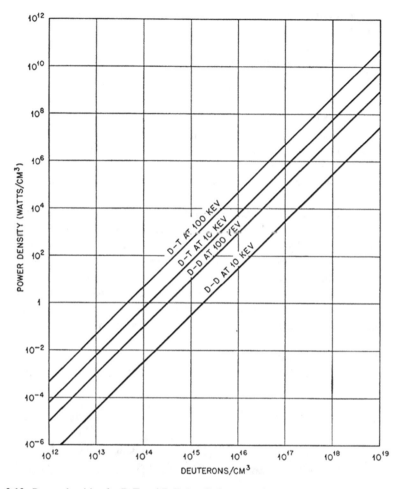

Fig. 2.12 Power densities for D-T and D-D (total) thermonuclear reactions

2.7   Critical Ignition Temperature for Fusion

The fusion temperature obtained by setting the average thermal energy equal to the Coulomb barrier gives too high temperature because fusion can be initiated by those particles which are out on the high-energy tail of the Maxwellian distribution of particle energies. The critical ignition temperature is lowered further by the fact that some particles, which have energies below the Coulomb barrier, can tunnel through the barrier.

The presumed height of the Coulomb barrier is based upon the distance at which the nuclear strong force could overcome the Coulomb repulsion. The required temperature may be overestimated if the classical radii of the nuclei are used for this distance, since the range of the strong interaction is significantly greater than a

classical proton radius. With all these considerations, the critical temperatures for
the two most important cases are about:

$$\text{Deuterium-Deuterium Fusion}: \ 40 \times 10^7 \text{K}$$

$$\text{Deuterium-Tritium Fusion}: \ 4.5 \times 10^7 \text{K}$$

The Tokamak Fusion Test Reactor (TFTR), for example, reached a temperature of
5.1×10^8 K, well above the critical ignition temperature for D-T fusion. TFTR was
the world's first magnetic fusion device to perform extensive scientific experiments
with plasmas composed of 50/50 deuterium/tritium (D-T), the fuel mix required for
practical fusion power production, and also the first to produce more than ten
million watts of fusion power.

The Tokamak Fusion Test Reactor (TFTR) was an experimental tokamak built at
Princeton Plasma Physics Laboratory (in Princeton, New Jersey) circa 1980. Following on from the Poloidal Diverter Experiment (PDX), and Princeton Large
Torus (PLT) devices, it was hoped that TFTR would finally achieve fusion energy
break-even. Unfortunately, the TFTR never achieved this goal. However, it did
produce major advances in confinement time and energy density, which ultimately
contributed to the knowledge base necessary to build the International Thermonuclear Experimental Reactor (ITER). TFTR operated from 1982 to 1997 (see
Fig. 2.13).

ITER is an international nuclear fusion research and engineering megaproject,
which will be the world's largest magnetic confinement plasma experiment. It is an

Fig. 2.13 Physical shape of TFTR in Princeton Plasma Physics Laboratory

Fig. 2.14 Sectional view of
ITER's tokamak

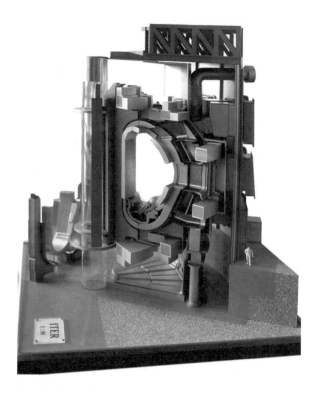

experimental tokamak nuclear fusion reactor, which is being built next to the
Cadarache facility in Saint-Paul-lès-Durance, south of France. Figure 2.14 is
depiction of the sectional view of ITER comparing to the man scale standing to
the lower right of the picture.

   In summary, temperature for fusion that required to overcome the Coulomb
barrier for fusion to occur is so high as to require extraordinary means for their
achievement:

$$\text{Deuterium-Deuterium Fusion}: 40 \times 10^7 \text{K}$$

$$\text{Deuterium-Tritium Fusion}: 4.5 \times 10^7 \text{K}$$

In the Sun, the proton-proton cycle of fusion is presumed to proceed at a much
lower temperature because of the extremely high density and high population of
particles:

$$\text{Interior of the sun, proton cycle}: 1.5 \times 10^7 \text{K}$$

2.8   Controlled Thermonuclear Ideal Ignition Temperature

The minimum operating temperature for a self-sustaining thermonuclear fusion reactor of magnetic confinement type (MCF) is that at which the energy deposited by nuclear fusion within the reacting system just exceeds that lost from the system as a result of Bremsstrahlung emission which is thoroughly described in the next two sections of this chapter here.

To determine its value, it is required to calculate the rates of thermonuclear energy production at a number of temperatures, utilizing Equations 2.47 and 2.50 together with Fig. 2.10, for *charged particle products only*, and to compare the results with the rates of energy loss as Bremsstrahlung derived from the following equations as Equations 2.53 and 2.54:

$$P_{DD(br)} = 5.5 \times 10^{-31} n_D^2 T_e^{1/2} \, W/cm^3 \tag{2.53}$$

and

$$P_{DT(br)} = 2.14 \times 10^{-30} n_D n_T T_e^{1/2} \, W/cm^3 \tag{2.54}$$

Note that the above two equations are established with the assumption that, for a plasma consisting only of hydrogen isotopes, $Z = 1$, and n_i and n_e are equal, so that the factor $n_e \sum (n_i Z^2)$ (this is described later in this chapter under Bremsstrahlung emission rate) may be replaced by n^2 where n is the particle density of either electrons or nuclei.

Note that the factor $n_e \sum (n_i Z^2)$ is sometimes written in the form $n_e^2 \left(\sum n_e Z^2 / \sum n_i Z \right)$, since n_e is equal to $\sum n_i Z$.

The assumption that we have made here and utilizing both Equations 2.53 and 2.54 arise from the fact that, in the plasma, the kinetic ion (nuclear) temperature and the electron temperature are the same.

To illustrate the ideal ignition temperature schematically, we take n_D to be as 10^{15} nuclei/cm^3 for the D-D reactions, whereas n_D and n_T are each 0.5×10^{15} nuclei/cm^3 for the D-T reaction. This makes Bremsstrahlung losses the same for the two cases. The results of the calculations are shown in Fig. 2.15.

The energy rates are expressed in terms of the respective power densities, i.e., energy produced or lost per unit time per unit volume of reaching system. It seems that the curve for the rate of energy is lost as Bremsstrahlung intersects the D-T and D-D energy production curves at the temperatures of 4 and 36 KeV, i.e., 4.6×10^7 and 4.1×10^8 K, respectively. These are sometimes called the *ideal ignition temperature*.

If we assume a Maxwellian distribution of electron velocities, then for rate of Bremsstrahlung energy emission per unit volume, it provides an accurate treatment and equation for total power radiation P_{br} as:

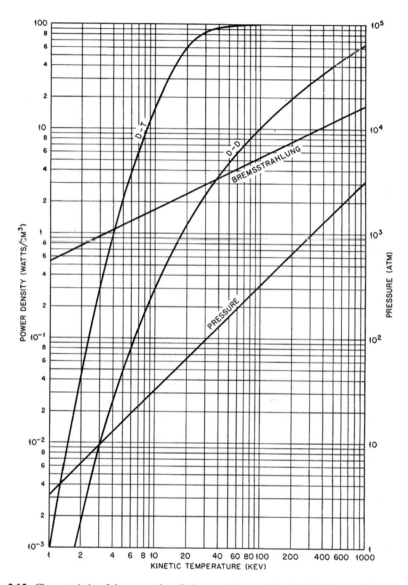

Fig. 2.15 Characteristic of thermonuclear fusion reactions and the ideal ignition temperature [1]

$$P_{\mathrm{br}} = g\,\frac{32\pi}{3\sqrt{3}} \cdot \frac{(2\pi kT)^{1/2} e^6}{m_{\mathrm{e}}^{3/2} c^3 h}\, n_{\mathrm{e}} \sum n_{\mathrm{i}} Z^2 \tag{2.55}$$

This equation will be explained later on, in more details, and then the ideal ignition temperature values defined above are the lowest possible operating temperatures for a self-sustaining thermonuclear fusion reactor. For temperatures lower than the

ideal ignition values, the Bremsstrahlung loss would exceed the rate of thermonuclear energy deposition by charged particles in the reacting system.

There exist two other factors, which require the actual plasma kinetic temperature to exceed the ideal ignition temperature values given above. These are in
addition to various losses besides just Bremsstrahlung radiation losses (Sect. 2.10)
that we possibly can be minimized, but not completely eliminate in a thermonuclear
fusion power plant reactor:

1. We have not yet considered the Bremsstrahlung emission as described later
   (Sect. 2.11), arising from Coulomb interaction of electrons with the helium
   nuclei produced in the thermonuclear fusion reactions as it is shown in
   Fig. 2.20. Since they carry two unit charges, the loss of energy will be greater
   than for the same concentration of hydrogen isotope ions.
2. At high temperatures present in a thermonuclear fusion reactions, the production
   of Bremsstrahlung due to electron-electron interactions is very distinctive than
   those resulting from the electron-ion interactions that are considered above. This
   is a concern, providing that the relativistic effects do not play in the game, and
   there should not be any electron-electron Bremsstrahlung, but at high electron
   velocities, such is not the case and appreciable losses can occur from this form of
   radiation.

In addition to power densities, Fig. 2.15 reveals the pressures at the various
temperature stages, based on the ideal gas equation $p = (n_i + n_e)kT$, where
$(n_i + n_e)$ is the total number of particles of nuclei and electron, respectively, per cm
^{3}, and T is the presentation of kinetic temperature in Kelvin. Under the present
condition here, $n_i = n_e = 10^{15}$ particles/cm^3, so that $(n_i + n_e) = 2 \times 10^{15}$.

With k having dimension of erg/K, the values are found in dimension of dynes/
cm^2, and the results have been converted to atmospheres assuming 1 atm = 1.01
$\times 10^6$ dynes/cm^2 and then plotted in Fig. 2.15. This figure also shows that the
thermonuclear power densities near the ideal ignition temperatures are in the
range of 100–1000 W/cm^3, which would be reasonable for continuous reactor
operation of a thermonuclear fusion reaction and that is the reason behind choosing
the density values as 1015 nuclei/cm^3 for the purpose of illustrating reacting
particles [1].

It should be noted that although the energy emitted as Bremsstrahlung may be
lost as far as maintaining the temperature of the thermonuclear reacting system is
concerned, it would not be a complete loss in the operating fusion reactor. Later on
in Sect. 2.11, we can demonstrate that the energy distribution of the electron
velocities is Maxwellian or approximately so and dependence of the Bremsstrahlung energy emission on the wavelength or photon energy and related equation can
be derived as well [1].

2.9   Bremsstrahlung Radiation

Bremsstrahlung is a German term that means "braking rays." It is an important phenomenon in the generation of X-rays. In the Bremsstrahlung process, a high-speed electron traveling in a material is slowed or completely stopped by the forces of any atom it encounters. As a high-speed electron approaches an atom, it will interact with the negative force from the electrons of the atom, and it may be slowed or completely stopped. If the electron is slowed down, it will exit the material with less energy. The law of conservation of energy tells us that this energy cannot be lost and must be absorbed by the atom or converted to another form of energy. The energy used to slow the electron is excessive to the atom and the energy will be radiated as X-radiation of equal energy. In summary, according to German dictionary "Bremsen" means to "break" and "Strahlung" means "radiation."

If the electron is completely stopped by the strong positive force of the nucleus, the radiated X-ray energy will have an energy equal to the total kinetic energy of the electron. This type of action occurs with very large and heavy nuclei materials. The new X-rays and liberated electrons will interact with matter in a similar fashion to produce more radiation at lower energy levels until finally all that is left is a mass of long wavelength electromagnetic wave forms that fall outside the X-ray spectrum.

Figure 2.16 here is showing Bremsstrahlung effect, produced by a high-energy electron deflected in the electric field of an atomic nucleus.

Characteristic of X-rays is an indication that they are emitted from heavy elements when their electrons make transition between the lower atomic energy levels. The characteristic X-ray emission which is shown as two sharp peaks in the illustration at the left occurs when vacancies are produced in the $n = 1$ or K-shell of the atom and electrons drop down from above to fill the gap. The X-rays produced

Fig. 2.16 Illustration of Bremsstrahlung effect

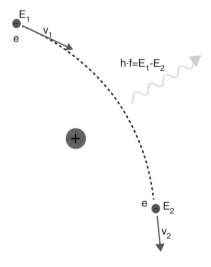

Fig. 2.17 X-ray
characteristic illustration

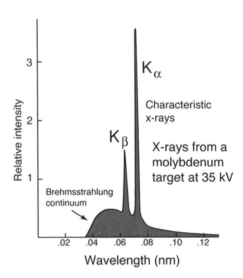

by transitions from $n = 2$ to $n = 1$ levels are called K-alpha X-rays, and those for the $n = 3 \rightarrow 1$ transition are called K-beta X-rays (see Fig. 2.17).

Transitions to the $n = 2$ or L-shell are designated as L X-rays, ($n = 3 \rightarrow 2$ are L-alpha, $n = 4 \rightarrow 2$ are L-beta, etc. The continuous distribution of X-rays which forms the base for the two sharp peaks at the left is called "Bremsstrahlung" radiation.

X-ray production typically involves bombarding a metal target in an X-ray tube with high-speed electrons which have been accelerated by tens to hundreds of kilovolts of potential. The bombarding electrons can eject electrons from the inner shells of the atoms of the metal target. Those vacancies will be quickly filled by electrons dropping down from higher levels, emitting X-rays with sharply defined frequencies associated with the difference between the atomic energy levels of the target atoms.

The frequencies of the characteristic X-rays can be predicted from the Bohr model. Moseley measured the frequencies of the characteristic X-rays from a large fraction of the elements of the periodic table and produced a plot of them, which is now called "Moseley's plot," and that plot is shown in Fig. 2.18 here for general knowledge purpose.

When the square root of the frequencies of the characteristic X-rays from the elements is plotted against the atomic number, a straight line is obtained. In his early 20s, Moseley measured and plotted the X-ray frequencies for about 40 of the elements of the periodic table. He showed that the K-alpha X-rays followed a straight line when the atomic number Z versus the square root of frequency was plotted. With the insights gained from the Bohr model, we can write his empirical relationship as follows:

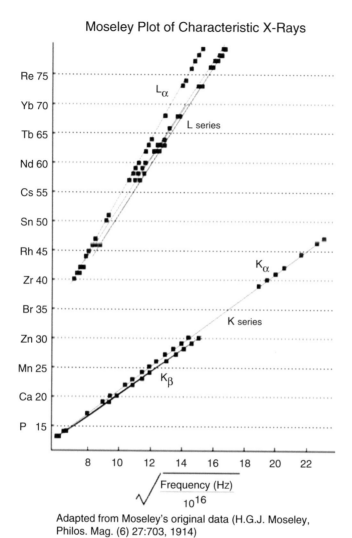

Fig. 2.18 Moseley's plot. Adapted from Moseley's original data (H. G. J. Moseley, Philos. Mag. (6) 27:703, 1914)

$$h\upsilon_{K_a} = 13.6\text{eV}(Z-1)^2\left[\frac{1}{1^2} - \frac{1}{2^2}\right] = \frac{3}{4}13.6(Z-1)^2\text{eV} \qquad (2.56)$$

Characteristic X-rays are used for the investigation of crystal structure by X-ray diffraction. Crystal lattice dimensions may be determined with the use of Bragg's law in a Bragg spectrometer.

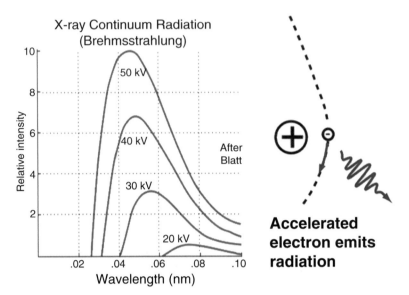

Fig. 2.19 Bremsstrahlung X-ray illustration

As it was stated above, "Bremsstrahlung" means "braking radiation" and is retained from the original German to describe the radiation, which is emitted when electrons are decelerated or "braked" when they are fired at a metal target. Accelerated charges give off electromagnetic radiation, and when the energy of the bombarding electrons is high enough, that radiation is in the X-ray region of the electromagnetic spectrum. It is characterized by a continuous distribution of radiation, which becomes more intense and shifts toward higher frequencies when the energy of the bombarding electrons is increased. The curves in Fig. 2.19 are from the 1918 data of Ulrey, who bombarded tungsten targets with electrons of four different energies.

The bombarding electrons can also eject electrons from the inner shells of the atoms of the metal target, and the quick filling of those vacancies by electrons dropping down from higher levels gives rise to sharply defined characteristic X-rays.

A charged particle accelerating in a vacuum radiates power, as described by the Larmor formula and its relativistic generalizations. Although the term Bremsstrahlung is usually reserved for charged particles accelerating in matter, not vacuum, the formulas are similar. In this respect, Bremsstrahlung differs from Cherenkov radiation, another kind of braking radiation which occurs only in matter and not in a vacuum.

The total radiation power in most established relativistic formula is given by:

$$P = \frac{q^2\gamma^4}{6\pi\varepsilon_0 c}\left[\dot{\beta}^2 + \frac{\left(\vec{\beta}\cdot\dot{\vec{\beta}}\right)^2}{1-\beta^2}\right] \tag{2.57}$$

where $\vec{\beta} = \vec{v}/c$ which is the ratio of the velocity of the particle divided by the speed of light and $\gamma = \frac{1}{\sqrt{1-\beta^2}}$ is the Lorentz factor, $\dot{\vec{\beta}}$. signifies a time derivation of $\vec{\beta}$, and q is the charge of the particle. This is commonly written in the mathematically equivalent form using as:

$$\left(\vec{\beta}\cdot\dot{\vec{\beta}}\right)^2 = \vec{\beta}^2\cdot\dot{\vec{\beta}}^2 - \left(\vec{\beta}\times\dot{\vec{\beta}}\right)^2 P = \frac{q^2\gamma^6}{6\pi\varepsilon_0 c}\left(\dot{\beta}^2 - \left(\vec{\beta}\times\dot{\vec{\beta}}\right)^2\right) \tag{2.58}$$

In the case where velocity of the particle is parallel to acceleration such as a linear motion situation, Equation 2.58 reduces to:

$$P_{a||v} = \frac{q^2 a^2\gamma^6}{6\pi\varepsilon_0 c^3} \tag{2.59}$$

where $a\equiv\dot{v} = \dot{\beta}c$ is the acceleration. For the case of acceleration perpendicular to the velocity $\left(\vec{\beta}\cdot\dot{\vec{\beta}} = 0\right)$, which is a case that arises in circular particle acceleration known as *synchrotron*, the total power radiated reduces to:

$$P_{a\perp v} = \frac{q^2 a^2\gamma^4}{6\pi\varepsilon_0 c^3} \tag{2.60}$$

The total power radiation in the two limiting cases is proportional to $\gamma^4(a\perp v)$ or $\gamma^6(a||v)$. Since $E = \lambda mc^2$, we see that the total radiated power goes as m^{-4} or m^{-6}, which accounts for why electrons lose energy to Bremsstrahlung radiation much more rapidly than heavier charged particles (e.g., muons, protons, alpha particles). This is the reason a TeV energy electron-positron collider (such as the proposed International Linear Collider) cannot use a circular tunnel (requiring constant acceleration), while a proton-proton collider (such as the Large Hadron Collider) can utilize a circular tunnel. The electrons lose energy due to Bremsstrahlung at a rate $(m_p/m_e)^4 \approx 10^3$ times higher than protons do.

As a general knowledge here, the nonrelativistic Bremsstrahlung formula for accelerated charges at a rate is given by the Larmor formula. For the electrostatic interaction of two charges, the radiation is most efficient, if one particle is an electron and the other particle is an ion. Therefore, Bremsstrahlung for the nonrelativistic case found the spectral radiation power per electron as:

$$P_{\mathrm{v}} = 2\pi P_\omega = \frac{\mathrm{d}E}{\mathrm{d}t\mathrm{d}\upsilon} = \frac{n_{\mathrm{i}}Z^2 e^6}{6\pi^2 \varepsilon_0^3 c^3 m_{\mathrm{e}}^2 \upsilon} \ln\left(\frac{b_{\max}}{b_{\min}}\right) \quad h\nu \ll m_{\mathrm{e}}\upsilon^2 \qquad (2.61)$$

where $b_{\max}$ and $b_{\min}$ are the maximum and minimum projectile to travel a distance of approximately b, respectively. This distance can be used for projectile impulse duration τ as $\tau = b/\upsilon_0$, where υ_0 is the incoming projectile velocity. Note that, on average, the impulse is perpendicular to the projectile velocity.

2.10 Bremsstrahlung Plasma Radiation Losses

Now that we have some understanding of the physics of Bremsstrahlung radiation, now we can pay our attention to *Bremsstrahlung plasma radiation losses*. So far our discussion has been referred to the energy or power that might be produced in a thermonuclear fusion reactor. This energy must be competed with inevitable losses, and the role of the processes which result in such losses is very crucial in determining the operating temperature of a thermonuclear reactor. Some energy losses can be minimized by a suitable choice of certain design parameters [1], but others are included in the reacting system that can be briefly studied and considered here.

Certainly Bremsstrahlung radiation from electron-ion and electron-neutral collisions can be expected. The radiation intensity outside the plasma region will be a function of various factors inside the plasma region such as the electron "kinetic temperature," the velocity distribution, the plasma opacity, the "emissivity," and the geometry. For example, in case of opacity, if we consider a mass of deuterium so large that it behaves as an optically thick or opaque body as far as Bremsstrahlung is concerned, and these radiations are essentially absorbed within the system. Under that assumption, then the energy loss will be given by the blackbody radiation corresponding to existing temperature. Note that even at ordinary temperatures, some D-D reactions will occur, although at an extremely slow rate.

The opacity and emissivity in the microwave region are determined by electron density and collision frequency, both measurable quantities. If strong magnetic field is present, the effects of gyro-resonance must also be accounted for in obtaining opacity [3].

Our understanding to date of the effects of non-Maxwellian velocity distributions on the radiation at microwave frequencies is not very complete. However, apparently, if the collision frequency is of the order of the viewing frequency, the actual velocity distribution is not very important because of the rapid randomization. For other cases, however, which in general are the ones of interest in this subject, there still remains much work to be done [3].

At kinetic temperatures in the region of 1 KeV or more, substances of low mass number are not only wholly vaporized and dissociated into atoms, but the latter are entirely stripped of their orbital electrons. In other words, matter is in a state of complete ionization; it consists of a gas composed of positively charged nuclei and an equivalent number of negative electrons, with no neutral particles. With this latter statement in hand, we can define the meaning of completely or fully ionized gas, which is a characteristic of plasma as well.

An ionized gaseous system consisting of equivalent numbers of positive ions and electrons, irrespective of whether neutral particles are present or not, is referred to as plasma, in addition to what was said in Chap. 1 for definition of plasma. At sufficiently high temperature, when there are no neutral particles and the ions consist of bare nuclei only, with no orbital electron, the plasma may be said to be completely or fully in ionized state.

We now turn our attention to plasma Bremsstrahlung radiation and to the principal source of radiation from fully ionized plasmas, Bremsstrahlung, with magnetic fields present, cyclotron or synchrotron radiation, as it was described in the previous section. The spectral range of Bremsstrahlung is very wide and extends from just above the plasma frequency into X-ray continuum for typical plasma range. By contrast, the cyclotron spectrum is characterized by line emission at low harmonics of the Larmor frequency. Similarly, synchrotron spectra from relativistic electron display distinctive characteristic [4].

Moreover, whereas cyclotron and synchrotron radiation can be dealt with classically, the dynamics being treated, from a relativistic viewpoint in the case of synchrotron radiation, Bremsstrahlung, from plasmas has to be interpreted from quantum mechanics perspective, through not usually relativistic. Bremsstrahlung radiation results from electrons undergoing transitions between two states of the continuum in the field of an ion or atom.

If the ions in plasma are not completely stripped, emission of energy will take place in the form of optical or excitation radiation. An electron attached to such an ion can absorb energy, e.g., as the result of a collision with a free electron, and thus be raised to an excited state. When the electron returns to a lower quantum level, the excitation energy is emitted in the form of radiation. This represents a possible source of energy loss from the plasma in a thermonuclear fusion reaction system that is considered as fusion reactor. Hydrogen isotope atoms have only a single electron and are completely stripped at a temperature of about 0.05 KeV, so that there is no excitation radiation above this temperature. However, if impurities of higher atomic number are present, energy losses in the form of excitation radiation can become very significant, especially at the lower temperature, while the plasma is being heated, and even at temperatures as high as 10 KeV [1].

If we ignore impurities in the plasma for the time being, we may state that the plasma in a thermonuclear fusion reactor system will consist of completely stripped nuclei of hydrogen isotope with an equal number of electrons at appropriate kinetic temperature. From such plasma, energy will inevitably be lost in the form of Bremsstrahlung radiation, that is, continuous radiation emitted by charge particle, mainly electrons, as a result of deflection by Coulomb fields of other charged particles. See Fig. 2.20, where in this figure b denotes the impact parameter and angle θ the scattering angle.

While beam energies below the Coulomb barrier prevent nuclear contributions to the excitation process, peripheral collisions have to select in the regime of

Fig. 2.20 Coulomb
scattering between an
electron and ion

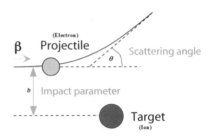

intermediate-energy Coulomb excitation to ensure the dominance of the electromagnetic interaction. This can be accomplished by restricting the analysis to events
at extremely forward scattering angles, corresponding to large impact parameters.

Except possibly at temperature about 50 KeV, the Bremsstrahlung from a
plasma arises almost entirely from electron-ion interactions as it is shown in
Fig. 2.19. Since the electron is free before its encounter with an ion and remains
free, subsequently, the transitions are often described as "free-free" absorption
phenomena, which also can be seen both in Inertial Confinement Fusion (ICF)
and Magnetic Confinement Fusion (MCF) thermonuclear reactions as well as it is
considered in inverse Bremsstrahlung effects, which is the subject of the next
section here.

In theory, the losses due to Bremsstrahlung could be described if the dimensions
of the system were larger than the mean free path for absorption of the radiation
photons under the existing conditions as it was described before. What these
conditions are telling us is that the system or magnetic fusion reactor would be
tremendously and impossibly large. This may end up with dimensions as large as 10
^{6} cm or roughly 600 miles or more, even at very high plasma densities. In a system
of this impractical size, a thermonuclear fusion reaction involving deuterium
(D) could become self-sustaining without the application of energy from the outside
source. In other words, a sufficiently large mass of deuterium could attain a critical
size, by the propagation of a large thermal chain reaction, just as does a suitable
mass of fissionable material as the result of a neutron chain reaction [1].

2.11 Bremsstrahlung Emission Rate

Using a classical expression for the rate P_c at which energy is radiated by an
accelerated electron, we can then write:

$$P_c = \frac{2e^2}{3c^3} a^2 \tag{2.62}$$

Fig. 2.21 Coulomb interaction of electron with a nucleus

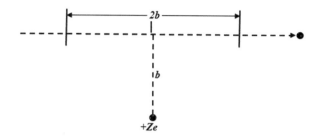

where

$e = $ Is the electric charge
$c = $ Is the velocity of light
$a = $ Is the particle acceleration

Per expression presented in Equation 2.62, we can also make an expression for the rate of electron-ion Bremsstrahlung energy emission of the correct but differing by a small numerical factor that may be obtained by a procedure that is more rigorous.

If we suppose an electron that moves past a relatively stationary ion of charge Ze with an impact parameter b as we saw in Fig. 2.20 and illustrated in Fig. 2.21 in different depiction as well.

The significance of impact parameter b can be defined in the absence of any electrostatic forces, which is the distance of closest approach between two particles. This will appear as an approximate value of large-angle, single-collision cross section for short-range interaction, or close encounter between charged particles may be obtained by a simple, classical mechanics and electromagnetic treatment based on Coulomb's law.

The magnitude of this distance will determine the angle of deflection of one particle by the other. Let, for a deflection of 90°, the impact parameter be b_0 as shown in Fig. 2.22 and by making a simplifying assumption that the mass of scattered particle is less than that of scattering particle so that the latter remains essentially stationary during this encounter, It is found from Coulomb's law that, for 90° deflections, the particles are a distance $2b_0$ apart at the point of closest approach.

From the viewpoint of classical electrodynamics, we see that the mutual potential Coulomb energy is equal to the center of mass or relative kinetic energy E of interacting particles. In the case of a *hydrogen isotope* plasma, all the particles carry the unit charge e, and the mutual potential energy at the point of closest approach is $e^2/2b_0$, and by law of conservation of energy, we can write:

$$E = \frac{e^2}{2b_0}$$
(2.63a)

or

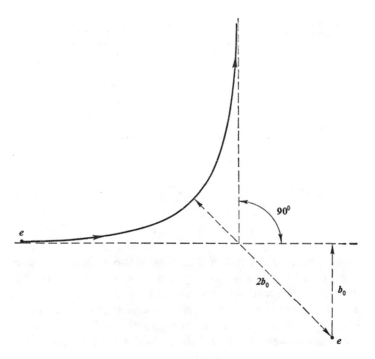

Fig. 2.22   Short-range Coulomb interaction for 90° deflections

$$b_0 = \frac{e^2}{2E} \qquad (2.63b)$$

Now if continue with beginning of this section and our concern about Bremsstrahlung emission rate, we go on to say that the Coulomb force between the charged particles is then Ze^2/b^2. Now let m_e to be the electron rest mass; then it its acceleration is $Ze^2/b^2 m_e$, and the rate of energy loss are as radiation is given by Equation 2.62 as:

$$P_e \approx \frac{2e^6 Z^2}{3m_e^2 c^3 b^4} \qquad (2.64)$$

If we designate the electron path length over which the Coulomb force is effective with $2b_0$ as it is illustrated in Fig. 2.21, and if the velocity is v, then the time during which acceleration occurs is $2b/v$. However, if the acceleration is assumed to be constant during this time, then the total energy E_e radiated as the electron moves past an ion with an impact parameter is written as:

$$E_e \approx \frac{4e^6 Z^2}{3 m_e^2 c^3 b^3 v} \tag{2.65}$$

Multiplying Equation 2.65 by n_e and n_i that are the numbers of electrons and ions, respectively, per unit volume, and also by velocity v, the result is the rate of energy loss P_a per unit impact area for all ion-electron collisions occurring in unit volume at an impact parameter b; then we can write:

$$P_a \approx \frac{4e^6 n_e n_i Z^2}{3 m_e^2 c^3 b^3} \tag{2.66}$$

The total power P_{br} radiated as Bremsstrahlung per unit volume is obtainable upon multiplying Equation 2.66 by $2\pi b db$ and integrating over all values of b from b_{min}, the distance of closest approach of an electron to an ion, to infinity, thus, the result would be as:

$$\begin{aligned}
P_{br} &\approx \frac{8\pi e^6 n_e n_i Z^2}{3 m_e^2 c^3 b_{min}} \int_{b_{min}}^{b} \frac{db}{b^2} \\
&= \frac{8\pi e^6 n_e n_i Z^2}{3 m_e^2 c^3 b_{min}}
\end{aligned} \tag{2.67}$$

An estimate of the minimum value of the impact parameter can be made by utilizing the Heisenberg uncertainty principle relationship, i.e.:

$$\Delta x \Delta p \approx \frac{h}{2\pi} \tag{2.68}$$

When Δx and Δp are the uncertainties in position and momentum, respectively, of a particle and h is Planck's constant, the uncertainty in the momentum may be set to the momentum $m_e v$ of the electron, and Δx may then be identified with b_{min}, so that:

$$b_{min} \approx \frac{h}{2\pi m_e v} \tag{2.69}$$

Furthermore, we assume a Maxwellian distribution of velocity among the electrons; it is possible to write:

$$\frac{1}{2} m_e v^2 = \frac{3}{2} k T_e \tag{2.70}$$

where T_e is the kinetic temperature of the electrons; hence,

$$b_{\min} \approx \frac{h}{2\pi(3kT_e m_e)^{1/2}} \tag{2.71}$$

Substituting Equation 2.71 into Equation 2.67, the result would be as:

$$P_{\mathrm{br}} \approx \frac{16\pi^2}{3^{1/2}} \cdot \frac{(kT_e)^{1/2} e^6}{m_e^{3/2} c^3 h} n_e n_i Z^2 \tag{2.72}$$

Equation 2.72 refers to a system containing a single ionic species of charge Z. In the case of a mixture of the ions or nuclei, it is obvious that the quantity $n_i Z^2$ should be replaced by $\sum (n_i Z^2)$, where the summation is taken over all the ion present. Note that the factor $n_e \sum (n_i Z^2)$ is sometime written in the form $n_e^3 \left(\sum n_e Z^3 / \sum n_i Z \right)_i$, since n_e is equal to $\sum n_i Z$. This was mentioned in Sect. 2.8 as well.

As we mentioned above and presented in Equation 2.55, as a more precise treatment, assume Maxwellian distribution of electron velocities gives for the rate of Bremsstrahlung energy emission per unit volume and write the same equation again:

$$P_{\mathrm{br}} = g \frac{32\pi}{3\sqrt{3}} \cdot \frac{(2\pi kT)^{1/2} e^6}{m_e^{3/2} c^3 h} n_e \sum n_i Z^2 \tag{2.73}$$

where g is the Gaunt factor which corrects the classical expression for the requirements of quantum mechanics. At high temperatures, the correction factor approaches a limiting value of $2 \times 3^{1/2}/\pi$, and taking this result, together with the known values of Boltzmann constant k in erg/K, e in statcoulombs, and m_e, c, and h in cgs units, Equation 2.73 or exact equation that is written as Equation 2.55 becomes:

$$P_{\mathrm{br}} = 1.57 \times 10^{-27} n_e \sum (n_i Z^2) T_e^{1/2} \mathrm{ergs}/(\mathrm{cm}^3)(s) \tag{2.74}$$

where T_e is the electron temperature in K, or making use of the conversion factor given as T_e, KeV is equivalent to $1.16 \times 10^7 T_e$ K, where 1 KeV $= 1.16 \times 10^7$ K.

The classical expression for the rate of Bremsstrahlung emission per unit volume frequency interval in the frequency range from v to $v + dv$ is given as:

$$dP_v = g \frac{32\pi}{3^{3/2}} \left(\frac{2\pi}{kT} \right)^{1/2} \frac{e^6}{m_e^{3/2}} \sum (n_i Z^2) \exp(-kv/kT) dv \, \mathrm{W/cm^3 (angstrom)} \tag{2.75}$$

If we integrate Equation 2.75 over all frequencies, this expression leads to either Equation 2.55 or 2.73, and for our purpose, it is more convenient to express Equation 2.75 in unit wave length in the interval from λ to $\lambda + d\lambda$ and that is:

$$dP_\lambda = 6.01 \times 10^{-30} g n_e \sum \left(n_i Z^2\right) T_e^{-1/2} \lambda^{-2} \exp(-12.40/\lambda T_e) d\lambda \qquad (2.76)$$

where the temperature is in kilo-electron volts (KeV) and the wavelengths are in angstrom.

If we assume the Gaunt factor g to remain constant, as is not strictly correct, the relative values of $dP_\lambda/d\lambda$ obtained from Equation 2.76, for arbitrary electron and ion densities, have been plotted as a function of wavelength as it can be seen in Fig. 2.23 for electron temperature of 1, 10, and 100 KeV. It can be observed that each curve passes through a maximum at a wavelength which differentiation of Equation 2.76 shows to be equal to $6.20/T_e$ angstroms.

Note that to the left of the maximum, the energy emission as Bremsstrahlung is dominated by the exponential term and decreases rapidly with decreasing wavelength. To the right of the maximum, however, the variation approaches a dependence upon $1/\lambda^2$, and the energy emission falls off more slowly with increasing wavelength of the radiation [1].

2.12   Additional Radiation Losses

As we briefly described in Sect. 2.8, in addition to various losses apart from Bremsstrahlung radiation loss, which can be minimized but not completely eliminated or contained in a practical reactor, there were two other factors, which were affecting such additional losses. To further enhance these concerns, and consider them for prevention of energy losses, we look at the following sources of energy losses.

According to Equation 2.73, the rate of energy loss as Bremsstrahlung increases with the ionic charge Z is equal to the atomic number in a fully ionized gas consisting only of nuclei and electrons. Consequently, the presence of impurities of moderate and high atomic number in thermonuclear reactor system will increase the energy loss, and as result the minimum kinetic temperature at which there is a net production of energy will also be increased [1].

However, if we consider a fully ionized plasma mixture containing n_1 nuclei/cm^3 of hydrogen isotopes ($Z = 1$) and n_e nuclei/cm^3 of an impurity of atomic number Z, then the electron density n_e is $n_1 + n_e Z$/cm^3. Thus, the factor $n_e \sum \left(n_i Z^2\right)$ in Equation 2.67 needs to be equal to $(n_1 + n_z Z)(n_1 + n_z Z^2)$. In the absence of the impurity, thus, the corresponding factor would be n_1^2. This follows that, from Equation 2.75, we can write:

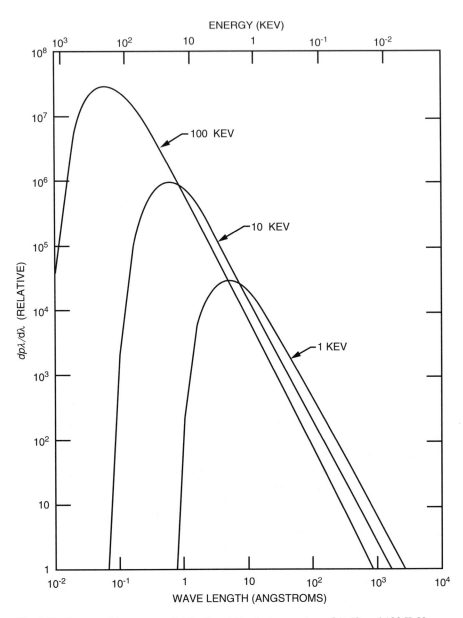

Fig. 2.23 Bremsstrahlung power distribution at kinetic temperature of 1, 10, and 100 KeV

$$\frac{\text{Power Loss in Presence of Impurity}}{\text{Power Loss in Presence of Impurity}} = \frac{(n_1 + z_z Z)(n_1 + z_z Z^2)}{n_1^2}$$

$$= 1 + f^2 Z^2 + fZ(Z + 1) \qquad (2.77)$$

where $f = n_e/n_1$, i.e., the fraction of impurity atoms.

Glasstone and Lovberg [1] argue that if the impurity is oxygen with atomic number $Z = 8$, and that is present to the extent of 1 at.%, so that $f = 0.01$, then in that case, Equation 2.77 results in the following value as:

$$\frac{\text{Power Loss in Presence of Impurity}}{\text{Power Loss in Presence of Impurity}} = 1.77 \qquad (2.78)$$

In other words what Equation 2.78 is telling us is that the presence of only 1 at.% of oxygen impurity will increase the rate of energy loss as Bremsstrahlung by 77 %. In the case of the D-D reaction system, Fig. 2.15 shows that this would raise the ideal ignition temperature from 36 to 80 KeV, and for D-T reaction, the same temperature increases from 4 to 4.5 KeV.

To remind again that "ideal ignition temperature is the minimum operation temperature for a self-sustaining thermonuclear reactor is that at which the energy deposited by nuclear fusion within the reacting system just exceeds that lost from the system as a result of bremsstrahlung emission" [1].

Per statement and example above, it is obvious for a thermonuclear reactor system with impurity of higher atomic number that the increase on ideal ignition temperature would be extremely high; thus, it appears to be an important requirement of a thermonuclear fusion reactor that even traces impurities, especially those of moderate and high atomic number. Therefore, such impurities should be rigorously excluded from the reacting plasma and there might be some possible exception to this rule [1].

To remind ourselves of an imperfect ionized impurity of plasma, we can also claim the following statement as well.

Imperfectly ionized impurity atoms with medium to high atomic number incur additional radiation losses in a plasma reactor. Electrons lose energy if these ions are further ionized or excited. This energy is then radiated from the plasma when later on an electron is captured; mainly recombination radiation or when ion returns to its original state, then radiation loss is via line radiation, respectively. Energy losses P_e^{LR} from both sources can be written in general form of:

$$P_e^{LR} = -n_e \sum_\sigma n_\sigma f_\sigma(T_e) \qquad (2.79)$$

where f_σ is a complicated function of T_e. Both line and recombination losses may exceed Bremsstrahlung losses by several orders of magnitude. As we talked about cyclotron effect in magnetic confinement of plasma, radiation from gyrating electrons also represents a loss source. Calculation of this one is very difficult in view of the fact that this radiation is partly reabsorbed in the plasma and partly reflected by the surrounding walls of the reactor. Fortunately, it is small compared with Bremsstrahlung losses under typical reactor conditions [5].

Note that recombination radiation is caused by *free-bound* transition. To elaborate further, we look at the final state of the electron that is a bound state of the atom or ion, if the ion was initially multiplied or ionized. The kinetic energy of the electron together with the difference in energy between the final quantum state n and the ionization energy of the atom or ion will appear as photon energy. This event involving electron capture is known as *radiative recombination* and emission as *recombination radiation*. In certain circumstances, recombination radiation may dominate over Bremsstrahlung radiation.

Other losses arise from energy exchange between components having different temperatures and from the interaction with the ever-present neutral gas background, namely, ionization and charge exchange. The study of these terms beyond the scope of this book and readers can refer to a textbook by Glasstone and Lovberg [1] as well as Raeder et al. [5].

2.13 Inverse Bremsstrahlung in Controlled Thermonuclear ICF and MCF

In case of laser-driven fusion, we have to be concerned by the dense plasma heating by inverse Bremsstrahlung, and it is very crucial for the design and critical evaluation of target for Inertial Confinement Fusion (ICF) to thoroughly understand the interaction of the laser radiation with dense, strongly coupled plasmas.

To accommodate the symmetry conditions needed, the absorption of laser energy must be carefully determined starting from the early stages [5, 6]. The absorption data for dense plasmas are also required for fast ignition by ultra-intense lasers due to creation of plasmas by the nanosecond pre-pulse [7]. Least understood are laser-plasma interactions that involve strongly coupled $\Gamma > 1$ and partially degenerate electrons. Such conditions also occur in warm dense matter experiments [8, 9] and laser-cluster interactions [10, 11].

The dominant absorption mechanism for lasers with the parameters typical for Inertial Confinement Fusion is inverse Bremsstrahlung. Dawson and Oberman [12] first investigated this problem for weak fields. Decker et al. [13] later extended their approach to arbitrary field strengths. However, due to the use of the classical kinetic theory, their results were inapplicable for dense, strongly coupled plasmas. This problem was addressed using a rigorous quantum kinetic description applying Green's function formalism [14, 15] or the quantum Vlasov approach [16]. However, these approaches are formulated in the high-frequency limit, which requires the number of electron-ion collisions per laser cycle to be relatively small. In the weak field limit, a linear response theory can be applied, and thus the strong electron-ion collisions were also included into a quantum description [17, 18] in this limit.

For dense strongly coupled plasmas, the approach for the evaluation of the laser absorption in both the high- and low-frequency limits must be fundamentally different. In the high-frequency limit, the electron-ion interaction has a collective rather than a binary character, and the laser energy is coupled into the plasmas via the induced polarization current. On the other hand, binary collisions dominate laser absorption in the low-frequency limit resulting in a Drude-like formulation.

At the intermediate frequencies, both strong binary collisions and collective phenomena have to be considered simultaneously. Interestingly, such conditions occur for moderate heating at the critical density of common Ny/Yag lasers.

Inverse Bremsstrahlung absorption in Inertial Confinement Fusion (ICF) or laser-driven fusion is an essential and fundamental mechanism for coupling laser energy to the plasma. Absorption of laser light at the ablation surface and critical surface of the pellet of D-T as target takes place via inverse Bremsstrahlung phenomenon in the following way:

• Laser intensity at the ablation surface causes the electrons to oscillate and consequently induced an electric field. The created energy due to the above oscillation of electrons will be converted into thermal energy via electron-ion collisions, which is known as inverse Bremsstrahlung process.

Bremsstrahlung and its inverse phenomena are linked in the following way:

• If two charged particles undergo a Coulomb collision as it was discussed before, they emit radiation, which is called again Bremsstrahlung radiation. Therefore, inverse Bremsstrahlung radiation is the opposite process, where electron scattered in the field of an ion absorbed a photon.

Note that b in Fig. 2.24 denotes the impact parameter that is defined as before and θ is the scattering angle.

Using the notation as given in Fig. 2.24, the differential cross section $d\sigma_{ei}/d\Omega$ for such a Coulomb collision is described by the Rutherford formula as follows:

$$\frac{d\sigma_{ei}}{d\Omega} = \frac{1}{4}\left(\frac{Ze^2}{m_e v^2}\right)^2 \frac{1}{\sin^4(\theta/2)} \tag{2.80}$$

where

$$\theta = \text{Is the scattering angle}$$
$$\Omega = \text{Is the differential solid angle}$$

Fig. 2.24 Coulomb scattering between an electron and ion

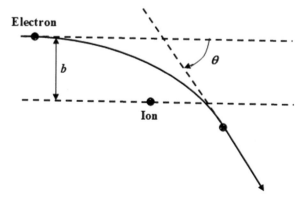

If we consider our analysis within spherical coordinate system, then the solid angle Ω is presented as:

$$d\Omega = 2\pi \sin\theta d\theta \tag{2.81}$$

In same coordinate, the impact parameter b is related to the scattering angle θ via the following formula as:

$$\tan\left(\frac{\theta}{2}\right) = \frac{Ze^2}{m_e v^2 b} \tag{2.82}$$

Substitution of Equations 2.81 and 2.82 along with utilization of Equation 2.80, we can now find the total cross section σ_{ei} for electron-ion collisions by integrating over all possible scattering angles and that is given as:

$$\sigma_{ei} = \int \frac{d\sigma_{ei}}{d\Omega} d\Omega = \frac{\pi}{2}\left(\frac{Ze^2}{m_e v^2}\right)^2 \int_0^{\pi} \frac{\sin\theta}{\sin^4(\theta/2)} d\theta \tag{2.83}$$

The integral from $\theta \to 0$ to $\theta \to \pi$, which is equivalent to $b \to \infty$ and $b \to 0$, diverges. However in plasma, the condition allows for us to define a lower and upper boundary limit b_{min} and b_{max}, respectively, and for that matter, the integration in Equation 2.83 reduces to the following form as:

$$\sigma_{ei} = \frac{\pi}{2}\left(\frac{Ze^2}{m_e v^2}\right)^2 \int_{b_{min}}^{b_{max}} \frac{\sin\theta}{\sin^4(\theta/2)} d\theta \tag{2.84}$$

The upper limit of this integral arises from Debye shielding that is defined in Chap. 1 of this book, which makes collision distance ineffective. Therefore, in a plasma the b_{max} limit can be replaced by Debye length λ_D. However, the lower limit b_{min} is often set to be equal to the Broglie wavelength, which Lifshitz and Pitaevskii [19] have shown that this approach is not adequate, and they derived the lower limit to be $b_{min} = Ze^2/k_B T_e$. Now that we have established the lower and upper bound limit, Equation 2.84 reduces to the following form in order to show the total cross section σ_{ei} in a plasma by:

$$\sigma_{ei} = \frac{\pi}{2}\left(\frac{Ze^2}{m_e v^2}\right)^2 \int_{Ze^2/k_B T_e}^{\lambda_D} \frac{\sin\theta}{\sin^4(\theta/2)} d\theta \tag{2.85}$$

Thus, having the knowledge of the cross section via Equation 2.85, one can calculate the collision frequency v_{ei} in the plasma. However, the collision frequency v_{ei} is defined as the number of collision and electron that undergoes with the background ions in plasma per unit time, and it depends on the ion density n_i, the cross section σ_{ei}, and the electron velocity v_e:

$$\nu_{ei} = n_i \sigma_{ei} \nu_e \tag{2.86}$$

In order to calculate the collision frequency ν_{ei}, we need to take the velocity distribution ν_e of the particles into account. In many cases it can be assumed that the ions are at rest $(T_i = 0)$ and electrons are in local thermal equilibrium. A Maxwellian electron velocity distribution, ν_e, in form of the following relation:

$$f(\nu_e) = \frac{1}{(2\pi k_B T_e/m)^{3/2}} \exp\left[-\left(\frac{m_e \nu_e^2}{2k_B T_e}\right)\right] \tag{2.87}$$

is isotropic and normalized in a way that:

$$\int_0^\infty \frac{1}{(2\pi k_B T_e/m)^{3/2}} \exp\left[-\left(\frac{m_e \nu_e^2}{2k_B T_e}\right)\right] = 1 \tag{2.88}$$

Using Equations 2.83 and 2.88 as well as performing the integrations, the electron-ion collision frequency results in:

$$\nu_{ei} = \left(\frac{2\pi}{m_e}\right)^{1/2} \frac{4Z^2 e^4 n_i}{3(k_B T_e)^{3/2}} \ln\Lambda \tag{2.89}$$

where $\Lambda = b_{max}/b_{min}$ and the factor Λ is called the Coulomb logarithm, a slowly varying term resulting from the integration over all scattering angles. In case of low-density plasmas and moderate laser intensities, driving the fusion reaction its value typically lies in the range of 10–20.

In order to derive Equation 2.89, the assumption was made on the fact that small-angle scattering events dominated, which is a valid assumption if the plasma density is not too high. For dense and cold plasma, Equation 2.89 is not applicable due to large-angle deflections becoming increasingly likely, violating the small-angle scattering assumption. If one uses the above stated method, the values of b_{min} and b_{max} can become comparable, so that ln Λ eventually turns negative, which is an obviously unphysical results.

In practical calculations a lower limit of ln $\Lambda = 2$ is often assumed; however, for dense plasmas a more complex treatment needs to be applied which is published by Bornath et al. [20] and Pfalzner and Gibbon [21].

Note that we need to be cautious if the laser intensity is very high, as in this case strong deviations from the Maxwell distribution can occur.

Readers can find more details in the book by Pfalzner [22].

Now that we have briefly analyzed the inverse Bremsstrahlung absorption for Inertial Confinement Fusion (ICF) or laser-driven fusion, we now pay our attention to this inverse event from physics of plasma point of view and consider the inverse Bremsstrahlung under free-free absorption conditions.

Free-free absorption inverse Bremsstrahlung takes place when an electron in continuum absorbs a photon. Its macroscopic equivalent is the collisional damping of electromagnetic waves. For a plasma in local thermal equilibrium, having found the Bremsstrahlung emission, we may then refer to Kirchhoff's law to find the free-free absorption coefficient α_ω. As we have stated before, the Bremsstrahlung emission coefficient is represented in terms of the Gaunt factor as an approximation in the form of:

$$\varepsilon_\omega(T_e) = \frac{8}{3\sqrt{3}} \frac{Z^2 n_e n_i}{m^2 c^3} \left(\frac{e^2}{4\pi\varepsilon_0}\right)^3 \left(\frac{m}{2\pi k_B T_e}\right)^{1/2} \bar{g}(\omega, T_e) e^{-\hbar\omega/k_B T_e} \qquad (2.90)$$

where $\bar{g}(\omega, T_e)$ is defined as:

$$\bar{g}(\omega, T_e) = \frac{\sqrt{3}}{\pi} \ln \left| \frac{2m}{\zeta\omega} \frac{4\pi\varepsilon_0}{Ze^2} \left(\frac{2k_B T_e}{\zeta m}\right)^{1/2} \right| \qquad (2.91)$$

From Equation 2.90, we can see that the Gaunt factor is a relatively slowly varying function of $\hbar\omega/k_B T_e$ over a wide range of parameters which means that the dependence of Bremsstrahlung emission on frequency and temperature is largely governed by the factor $(m/2\pi k_B T_e)^{1/2} \exp(-\hbar\omega/k_B T_e)$ in Equation 2.90. As it was also stated for laboratory plasmas with electron temperatures in the KeV range, the Bremsstrahlung spectrum extends into the X-ray region of the spectrum. Note that the factor $\sqrt{3}/\pi$ in Equation 2.91 is to conform with the conventional definition of the Gaunt in the quantum mechanical treatment.

In terms of the Rayleigh-Jeans limit, this gives a relationship for free-free absorption coefficient as follows:

$$\alpha_\omega(T_e) = \frac{64\pi^4}{3\sqrt{3}} \frac{Z^2 n_e n_i}{m^2 c\omega^2} \left(\frac{e^2}{4\pi\varepsilon_0}\right) \left(\frac{m}{2\pi k_B T_e}\right)^{1/2} \bar{g}(\omega, T_e) \qquad (2.92)$$

In Equation 2.91 $\zeta = 0.577$ is Euler's constant, and the factor $(2/\zeta) \simeq 1.12$ in the argument of the logarithm has been included to make $\bar{g}(\omega, T_e)$ in Equation 2.91 to agree with the exact low-frequency limit determined from the plasma Bremsstrahlung spectrum. In classical picture of plasma Bremsstrahlung spectrum, an exact classical treatment of an electron moving in the Coulomb field of an ion is a standard problem in classical electrodynamics. Provided the energy radiated as Bremsstrahlung is a negligibly small fraction of the electron energy where the ion is treated as a stationary target, then the electron orbit is hyperbolic, and the power spectrum $dp(\omega)/d\omega$ from a test electron colliding with plasma ions of density n_i may be written as:

$$\frac{dp(\omega)}{d\omega} = \frac{16\pi}{3\sqrt{3}} \frac{Z^2 n_i}{m^2 c^3} \left(\frac{e^2}{4\pi\varepsilon_0}\right)^3 \frac{1}{v} G(\omega b_0/v) \qquad (2.93)$$

where $b_0 = Ze^2/4\pi\varepsilon_0 m v^2$ is the impact parameter for $90°$ scattering, v the incident velocity of the electron, and $G(\omega b_2/v)$ is a dimensionless factor that is known as Gaunt factor as it was defined before, which varies only weakly with plasma frequency ω.

It can be shown that the dispersion relation for electromagnetic waves in an isotropic plasma becomes:

$$\frac{c^2 k^2}{\omega^2} = 1 - \frac{\omega_p^2}{\omega(\omega - iv_{ei})} \tag{2.94}$$

This is allowable phenomenologically for the effects of electron-ion collisions through a collision frequency v_{ei}. Further on, it can be shown that electromagnetic waves are damped as a result of electron-ion collisions, with damping coefficient $\gamma = v_{ei}\left(\omega_p^2/2\omega^2\right)$.

If we take Equation 2.94 into consideration, which is expressing the collision damping of electromagnetic waves and using this to obtain the absorption coefficient, we provided in Equation 2.90, with the Coulomb logarithm in place of the Maxwell-averaged Gaunt factor, a difference that reflects the distinction between these separate approaches. Whereas inverse Bremsstrahlung is identified with incoherent absorption of photon by thermal electrons, the result in Equation 2.94 is macroscopic in that it derives from a transport coefficient, namely, the plasma conductivity [23].

At the macroscopic level, electron momentum is driven by an electromagnetic field before being dissipated by means of collisions with ions. However, absorption of radiation by inverse Bremsstrahlung as expressed in Equation 2.92 is more effective at high densities and low electron temperature and for low-frequency plasmas. For the efficient absorption of laser light by plasma at the ablation surface of target pellet of D-T, the mechanism of the process is very important. We anticipate absorption to be strongest in the region of the critical density n_c, since this is the highest density to which incident light can penetrate. In the vicinity of the critical density $Zn_e n_i \tilde{} N_c^2 = (m\varepsilon_0/e^2)^2 \omega_L^4$, where ω_L is presenting the frequency of the laser light, so that free-free absorption is sensitive to the wavelength of the incident laser light [23].

References

1. S. Glasstone, R.H. Lovberg, *Controlled Thermonuclear Reactions* (D Van Nostrand Company, Inc, New York, NY, 1960)
2. B. Zohuri, P. McDaniel, *Thermodynamics in nuclear power plant systems* (Springer, New York, NY, 2015)
3. J.E. Drummond, *Plasma Physics* (Dover Publication, Mineola, NY, 2013)
4. T.J.M. Boyed, J.J. Sanderson, *The Physics of Plasmas* (Cambridge University Press, Cambridge, 2003)

5. J.D. Lindl et al., Phys. Plasmas **11**, 339 (2004)
6. M. Michel et al., Phys. Rev. Lett. **102**, 025004 (2009)
7. M.H. Key, Phys. Plasmas **14**, 055502 (2007)
8. A.L. Kritcher et al., Science **322**, 69 (2008)
9. E. Garcia Saiz et al., Nat. Phys. **4**, 940 (2008)
10. B.F. Murphy et al., Phys. Rev. Lett. **101**, 203401 (2008)
11. T. Bornath et al., Laser Phys. **17**, 591 (2007)
12. J.M. Dawson, C. Oberman, Phys. Fluids **5**, 517 (1962)
13. C.D. Decker et al., Phys. Plasmas **1**, 4043 (1994)
14. D. Kremp et al., Phys. Rev. E **60**, 4725 (1999)
15. T. Bornath et al., Phys. Rev. E **64**, 026414 (2001)
16. H.J. Kull, L. Plagne, Phys. Plasmas **8**, 5244 (2001)
17. H. Reinholz et al., Phys. Rev. E **62**, 5648 (2000)
18. A. Wierling et al., Phys. Plasmas **8**, 3810 (2001)
19. E.M. Lifshitz, L.P. Pitaevskii, *Physical Kinetics* (Pergamon Press, Oxford, 1981)
20. T. Bornath, M. Schlanges, P. Hillse, D. Kremp, Phys. Rev. E **64**, 026414 (2001)
21. S. Pfalzner, P. Gibbon, Phys. Rev. E **57**, 4698 (1998)
22. S. Pfalzner, *An Introduction to Inertial Confinement Fusion* (Taylor and Francis, Oxfordshire, 2006)
23. T.J.M. Boyd, J.J. Sanderson, *The Physics of Plasmas* (Cambridge University Press, Cambridge, 2005)

Chapter 3
Confinement Systems for Controlled Thermonuclear Fusion

An Increase on energy demands in our today's life going forward to the future has forced us to look into alternative production of energy in a clean way, along the nuclear fission and fossil fuel way of producing energy. Scientist are suggesting controlled thermonuclear fusion reaction as an alternative way of generating energy, either via magnetic confinement or inertial confinement of plasma to generate heat for producing steam and as result electricity to meet such increase on energy demand. Each of these approaches has their own technical and scientific challenges, which scientists need to overcome. This chapter talks about a way of confining plasma and the systems of the confinement, which are able to impose a controlled way of thermonuclear fusion reaction for this purpose.

3.1 Introduction

Fusion power is the generation of energy by nuclear fusion. Fusion reactions are high-energy reactions in which two lighter atomic nuclei fuse to form a heavier nucleus. This major area of plasma physics research is concerned with harnessing this reaction as a source of large-scale sustainable energy. There is no question of fusion's scientific feasibility, since stellar nucleosynthesis is the process in which stars transmute matter into energy emitted as radiation. Conversion of mass of matter to energy is very well understood and demonstrated by Einstein's theory of relativity and his famous formula as below:

$$E = MC^2 \tag{3.1}$$

where E is the kinetic energy produced by M, which is the reduced mass of two individual particles interacting with each other, and it was expressed by Equation 2.23 and multiplied by C that is the speed of light. Figure 3.1 is the presentation of

© Springer International Publishing AG 2016
B. Zohuri, *Plasma Physics and Controlled Thermonuclear Reactions Driven Fusion Energy*, DOI 10.1007/978-3-319-47310-9_3

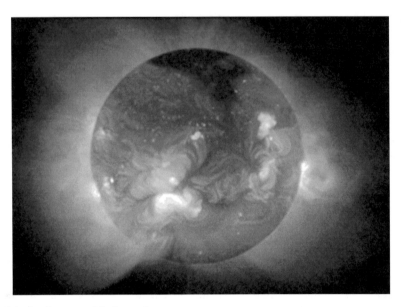

Fig. 3.1 Sun is a natural fusion reactor

such energy that is taking place at the surface of the Sun, in our solar system, which is a natural fusion reactor.

For reduced mass M to exist and relationship in Equation 3.1 to take place, the particles must come within range of the nuclear forces and surpass the Coulomb barrier via driven kinetic energy available in the center of the mass system of the colliding particles. As we observed in Chap. 2, it was realized that bombardment of light element targets with high-energy particle beams could not sufficiently produce enough power, unless the energy, necessarily imparted to outer shell electrons in the collision process, was utilized.

What the preceding text implies is that the reacting particles must be confined at high density for a time sufficiently long for energy transfer to the nuclei; this process is called the "break-even" condition, also known as the Lawson criterion [1] to take place.

The Lawson criterion is an important general measure of a system that defines the conditions needed for a fusion reactor to reach what is known as *ignition temperature*, which is the heating of plasma by the products of the fusion reactions to be sufficient to maintain the temperature of the plasma against all losses with external power input. As it was originally formulated in Chap. 2, the Lawson criterion gives a minimum required value for the product of the particle plasma density such as electron n_e and the *energy confinement time* τ_E.

Figure 3.2 is showing a typical Lawson criterion, or minimum value of electron density multiplied by energy confinement time required for self-heating, for their fusion reactions. For D-T reaction, $n_e\tau_E$ minimizes near the temperature 25 KeV or roughly 300 million Kelvin, as it can be seen in the figure.

Fig. 3.2 Depiction of the Lawson criterion for three fusion reactions

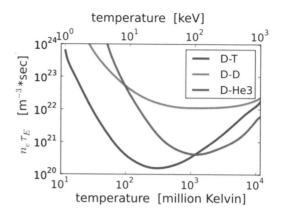

Note that, although the above text was argued based on magnetic confinement fusion (MCF) approach, similar reasoning would apply to inertial confinement fusion (ICF) by multiplying the density of plasma particles with the radius of the pellet containing the D-T for fusion reaction, which is shown further down in this chapter.

To summarize what we have discussed so far, we can express the following statement. At the temperature the reaction rate is taking place, it is proportional to the square of the density; the time during which confinement can be secured turns out to be limited to a small fraction of second, and therefore the density needed in order to achieve a useful power output is very high. See Sect. 2.5 as well.

In addition, the temperature required for barrier penetration and the density required (see Sect. 2.2) for a practical device will be determined from data concerning reaction cross sections, and they are representing conditions of matter known to exist in terrestrial galaxy surrounding us. The concept behind such phenomena on the Earth was first produced in technology of thermonuclear weapons which humankind realized, and similar conditions were used for triggering the most devastating weapons that are known to a human being.

Although, the first release of man-made thermonuclear energy via H-bomb took place in 1952, but the problem of how to control this sudden release in a controlled way, for the purpose of generating electric power, is still with us today.

3.2   Magnetic Confinement

The major magnetic fusion concepts that are in consideration by folks that are in quest of confining plasma for magnetic fusion concepts are:

1. The tokamak
2. The stellarator
3. The reversed-field pinch

4. The spheromak
5. The field-reversed configuration
6. The levitated dipole

All these magnetic fusion concepts except the stellarator are 2D axisymmetric toroidal configurations. However, the stellarator is an inherently 3D configuration, and we are just going to discuss here, in this section, the tokamak and stellarator configurations later on. We will discuss these two concepts, primarily from the point of view of macroscopic magnetohydrodynamics (MHD) equilibrium and stability.

Magnetic confinement of plasma is an attempt to prevent particles of moderate density around 10^{14}–10^{15} cm^{-3} in plasma to escape the reaction volume by thermal velocity for long periods (i.e., $\tau \geq 1$ second). The concept is based on the foundation that charged the particle path generally forming a spiral along magnetic field lines, which is created by the Lorentz force acting on plasma particle with charge q and moving with velocity of $\vec{v}$ in a magnetic field with induction of $\vec{B}$ as it was explained in Chap. 2.

The above approach is based on a single particle and its motion, depending on the density of charged particles of plasma and their behavior; they present a fluid, either with collective effects being dominant or as collective individual particles. In dense plasmas, the electrical forces between particles couple them to each other and to the electromagnetic fields, which affects their motions.

To have a better concept for single-particle approach and what does that mean, we look at the rarefied plasmas. Under these circumstances, the charged particles do not interact with one another, and their motions do not govern a large enough current to significantly affect the electromagnetic fields. Therefore, under these conditions, the motion of each particle, classically, can be treated independent of any other, by solving the Lorentz force equation for prescribed electric and magnetic field. This procedure is known as a single-particle approach and is valid for investigating high-energy particles in the Earth's radiation belts (i.e., Van Allen radiation belt), the solar corona, and in practical devices such as cathode ray tubes or traveling-wave amplifiers, which are few examples that could be mentioned. Figure 3.3 is an artistic concept of Van Allen belt cross section, which is an imaginary belt of radiation layer of energetic charged particles that are held in place around a magnetized plant, such as the Earth, by the planet's magnetic field.

As it sounds, in magnetized plasmas under the influence of an external static force or slowly varying magnetic field produced by the electric field, the single-particle approach is the only applicable classical solution for studying the charged particle motion utilizing the Lorentz force equation, which in general is defined as:

$$\vec{F} = m\vec{a} = m \cdot \frac{d\vec{v}}{dt} = q(\vec{E} + \vec{v} \times \vec{B}) \tag{3.2}$$

Equation 3.2 for motion of the charged particle in magnetized plasmas holds, if the external magnetic field is quite strong, compared to the magnetic field produced by

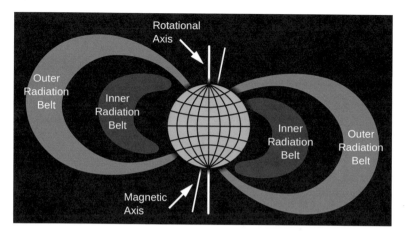

Fig. 3.3 Conceptual cross section of the Van Allen belt around the Earth

the electric current arising from the charged particle motions, an event that is very understandable by physics or theory of electromagnetism.

Note that here we are only concerned with nonrelativistic motion of charged particles that are obeying Newtonian classical mechanics rules and the second law of motion. Equation 3.2 is valid for the relativistic case, if we simply replace particle mass m with the famous Einstein formula of relativity in terms of $m = m_0$ $(1 - v^2/C^2)^{-1/2}$ where m_0 is the rest mass of particle. More commonly, the relativistic form of Equation 3.2 is written in terms of particle momentum $\vec{P} = m$ $\vec{v}$ rather than velocity $\vec{v}$.

CASE I: Uniform $\vec{E}$ and $\vec{B}$ Fields and $\vec{E} = 0$

As it was stated above for the simples cases of motion in uniform field, when a particle is under the domination of a static electric filed, which is uniform in space, the Lorentz force $\vec{F}$ is expressed in the following form with only a static and uniform magnetic field present:

$$\vec{F} = q\vec{v} \times \vec{B} \tag{3.3}$$

In this case, the particle moves with a constant acceleration along the direction of the field and does not warrant further study.

From the classical mechanics point of view, Lorentz force also is equal to the mass of the particle of interest multiplied by the mass of it, so we can write

$$\vec{F} = m\vec{a} = m \cdot \frac{d\vec{v}}{dt} \tag{3.4}$$

Combining the Equations 3.3 and 3.4, we can write the momentum balance equation for this type of particle is as:

$$m \cdot \frac{d\vec{v}}{dt} = q\vec{v} \times \vec{B} \tag{3.5}$$

For further analysis, we can decompose the particle velocity vector $\vec{v}$ into its two components, namely, parallel $\vec{v}_{||}$ and $\vec{v}_{\perp}$ perpendicular, respectively, to the magnetic field, i.e.,

$$\vec{v} = \vec{v}_{||} + \vec{v}_{\perp} \tag{3.6}$$

Lorentz force $\vec{F}$ is proportional to the vector product $\vec{v} \times \vec{B}$, it is vertical to the plane of vector velocity $\vec{v}$ and magnetic $\vec{B}$, and $\vec{v} \times \vec{B} = \vec{v}_{\perp} \times \vec{B}$ it is a function only of the velocity component $\vec{v}_{\perp}$ which is vertical to $\vec{B}$. Note that $\vec{v}_{\perp}$ is the vertical component of vector velocity $\vec{v}$. As far as parallel component of velocity is concerned, it has no effect, because $\vec{v}_{||} \times \vec{B} = 0$ is the component $\vec{v}_{||}$ of the particle velocity parallel to $\vec{B}$ and does not lead to any force influencing on the particle.

Using our knowledge of vector analyses and taking the dot product of Equation 3.5 with vector $\vec{v}$, we have:

$$\vec{v} \cdot m\frac{d\vec{v}}{dt} = \vec{v} \cdot q(\vec{v} \times \vec{B})$$
$$m\frac{1}{2}\left\{\frac{d(\vec{v} \cdot \vec{v})}{dt}\right\} = q[\vec{v} \cdot (\vec{v} \times \vec{B})] \tag{3.7}$$
$$\frac{d}{dt}\left(\frac{mv^2}{2}\right) = 0$$

where $v = |\vec{v}|$ is the speed of particle and as we have noted before, $(\vec{v} \times \vec{B})$ is perpendicular to $\vec{v}$ so the right-hand side is zero.

Obviously, from the above, we can see that the static magnetic field cannot change the kinetic energy of the particle, since the force is always perpendicular to the direction of motion, and this is true even for a spatially nonuniform field. This is because the deviation above did not use the fact that the field is uniform in space.

Using Equation 3.6 and rewriting Equation 3.5, we have:

$$\frac{d\vec{v}_{||}}{dt} + \frac{d\vec{v}_{\perp}}{dt} = \frac{q}{m}(\vec{v}_{\perp} \times \vec{B}) \tag{3.8}$$

However, as stated above, the term $\vec{v}_{||} \times \vec{B} = 0$ in Equation 3.8 can be split into two equations in terms of $\vec{v}_{||}$ and $\vec{v}_{\perp}$, respectively; thus, we have:

$$\frac{d\vec{v}_{||}}{dt} = 0 \rightarrow \vec{v}_{||} = \text{constant}$$

$$\frac{d\vec{v}_{\perp}}{dt} = \frac{q}{m}\left(\vec{v}_{\perp} \times \vec{B}\right)$$

(3.9)

Further investigation of Equation 3.9 reveals that the magnetic field $\vec{B}$ has no effect on the motion of the particle in the direction along it and that it only affects the particle velocity in the direction perpendicular to it.

We now consider a static magnetic field oriented along the z-axis in vector form as $\vec{B} = \hat{z}B$ in order to be able to examine the characteristic of the perpendicular further on. We can then write Equation 3.5 in component form as:

$$m\frac{dv_x}{dt} = qBv_y \tag{3.10a}$$

$$m\frac{dv_y}{dt} = -qBv_x \tag{3.10b}$$

$$m\frac{dv_z}{dt} = 0 \tag{3.10c}$$

The parallel component of particle velocity $\vec{v}_{||}$ to magnetic field is usually denoted as v_z and is constant, since the Lorentz force $q\left(\vec{v} \times \vec{B}\right)$ is perpendicular to $\}\backslash hat\{z\}\{$. To determine the time variations of v_x and v_y, we refer to Equations 3.10a and 3.10b by taking the second derivative of these equations in respect to time t to obtain the following sets of equations:

$$\frac{d^2 v_x}{dt^2} + \omega_c^2 v_x = 0 \tag{3.11a}$$

$$\frac{d^2 v_y}{dt^2} + \omega_c^2 v_y = 0 \tag{3.11b}$$

where $\omega_c = -qB/m$ is the *gyrofrequency* or *cyclotron frequency*, and we show it as the following equation:

$$\text{Cyclotron Frequency} \quad \boxed{\omega_c \equiv -\frac{qB}{m}} \tag{3.12}$$

The dimension of ω_c as an angular frequency is rad/m and can be a positive or negative value which is driven by the sign of charge q.

Figure 3.4 is the presentation of cylindrical coordinate with the azimuthal angle of ϕ with a right-hand sense of rotation along the positive direction from the x-axis, and the same picture shows the motion of particle as well, where z-axis is an indication of coming out of the page with the symbol of $\otimes$.

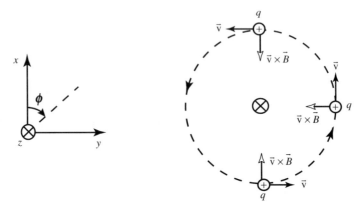

Fig. 3.4 Motion of particle in a magnetic field

The solution to linear differential sets of Equations 3.11a and 3.11b in the form of harmonic motion is provided as follows, assuming that $\vec{v}_\perp = v_\perp$ and $\vec{v}_{||} = v_z = v_{||}$:

$$v_x = v_\perp \cos\left(\omega_c + \psi\right) = v_\perp e^{i\omega_c t} = \frac{dx}{dt} = \dot{x} \tag{3.13a}$$

$$v_y = v_\perp \cos\left(\omega_c + \psi\right) = \frac{m}{qB}\dot{v}_x = \pm\frac{1}{\omega_c}\dot{v}_x = \pm i v_\perp e^{i\omega_c t} = \dot{y} \tag{3.13b}$$

$$v_z = v_{||} \tag{3.13c}$$

where ψ is some arbitrary phase angle, which defines the orientation of the particle velocity at $t = 0$, and $v_\perp = \sqrt{v_x^2 + v_y^2}$ is the constant speed in the plane perpendicular to the magnetic field $\vec{B}$.

Considering Fig. 3.4 and assuming a positive charge q in motion, at a different point along its orbit, we can clearly see that the particle experiences a force of $\vec{F} \propto \vec{v} \times \vec{B}$ directed inward at all times at any given points, which balances the centrifugal force, driven by the circular motion of the particle. For a magnetic field in the z-direction, in case of electron, the particle rotation follows the right-hand thump rule in electromagnetism.

The right-hand rule for magnetic force, describing the interactions between the current, the flow of electrons, and magnets, can be used to do useful work, like power motors, and will continue to be important in the future because they can be used for things like wireless energy transfer. This simple demonstration will show how strongly and quickly they interact with each other (see Fig. 3.5).

A more complicated right-hand rule (RHR) is Fleming's RHR, which describes the motion or force in which something moves. It is useful for understanding the direction of various players in electromagnetism, since they interact at right angles.

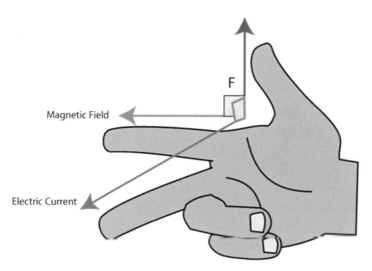

Fig. 3.5 The right-hand rule for magnetic force

The direction of the thumb is the direction of the force, the direction of the index finger indicates the direction of the magnetic field, and the direction of the middle finger is the direction of the electric current.

From what we have so far, we can easily find the radius of circular trajectory which can be found by considering the fact that the $\vec{v} \times \vec{B}$ force is balanced by the centripetal force; therefore, we have:

$$-\frac{mv_\perp^2}{r} = q\vec{v} \times \vec{B} = qv_\perp B \tag{3.14}$$

Using what we have for Equation 3.12 and substituting it into Equation 3.14, we get the result for the final form of trajectory radius of gyroradius, which also known as the *Larmor* radius and is written as:

$$r_c = \frac{-mv_\perp}{qB} = \frac{v_\perp}{\omega_c} \tag{3.15}$$

Note that the magnitude of the particle velocity remains constant, since the magnetic field force is at all times perpendicular to the motion as it can be seen in Fig. 3.4. Additionally, by the convention, the gyroradius is written in r_c and can take negative value. This is a mathematical formulation that allows for writing the expression for particle trajectory for either positive or negative charges in compact form. The gyroradius should always be interpreted as a real physical distance [2].

Note that the magnetic field has no influence over changing the kinetic energy of the particle; however, it does change the direction of its momentum. It is important to note that the gyrofrequency ω_c of the charged particle does not depend on its velocity or kinetic energy and is only a function of intensity of the magnetic field.

Further analyses can be done to show the particle position as a function of time by integration of Equations 3.13a and 3.13b sets to find the following information:

$$x = r_c \sin(\omega_c t + \psi) + (x_0 - r_c \sin\psi) \tag{3.16a}$$

$$y = -r_c \cos(\omega_c t + \psi) + (y_0 - r_c \cos\psi) \tag{3.16b}$$

$$z = z_0 + v_{||}t \tag{3.16c}$$

where x_0, y_0, and z_0 are the coordinates of the location of the particle at $t = 0$, and ψ is simply the phase with respect to a particular time of origin.

Plotting the trajectory function of sets of Equations 3.16a, 3.16b, and 3.16c shows that the particle moves in a circular orbit perpendicular to the magnetic field $\vec{B}$ with an angular frequency ω_c and radius r_c about a *guiding center*

$$\vec{r}_g = x_0\hat{x} + y_0\hat{y} + \left(z_0 + v_{||}t\right)\hat{z}.$$

If we are considering particle motion (i.e., electron) in inhomogeneous field, then the concept of a guiding center makes it very useful, since the gyration is often much more rapid than the motion of the guiding center. Now considering the sets of Equations 3.13a, 3.13b, and 3.13c, in their present form, influences the guiding center to simply move linearly along the z-axis at a uniform speed $v_{||}$ as it is depicted in Fig. 3.6, although the particle motion itself is helical.

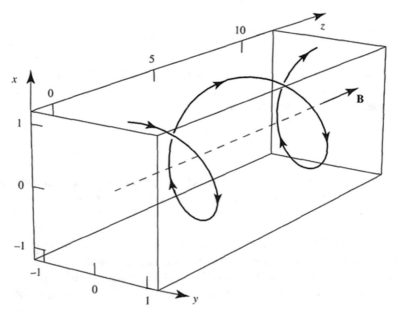

Fig. 3.6 The electron guiding center motion in a magnetic field $\vec{B} = B\hat{z}$ (Courtesy of Inan and Golkowski) [2]

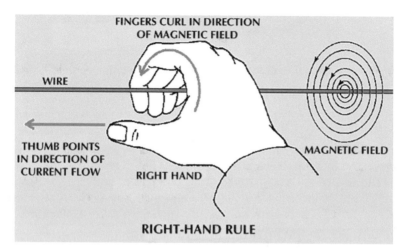

Fig. 3.7 Right-hand rule direction

From Fig. 3.6, the *pitch angle* of the helix is defined as:

$$\alpha = \tan^{-1}\left(\frac{v_\perp}{v_\parallel}\right) \tag{3.17}$$

Noticeably for both positive and negative charges such as proton or electron, respectively, the particle gyration constitutes an electric current in the $-\phi$ direction (i.e., opposite to the direction of the figures of the right hand when the thump points in the direction of the $+z$ axis). The conceptual direction using the right-hand rule is depicted in Fig. 3.7 here, and in that case, magnetic moment μ associated with such a current loop is given by current multiplied by area or mathematically presented as:

$$\mu = \underbrace{\left(\left|\frac{q\omega_c}{2\pi}\right|\right)}_{\text{current}} \underbrace{\left(\pi r_c^2\right)}_{\text{area}} = \frac{mv_\perp^2}{2B} \tag{3.18}$$

Similarly, if we are interested about the torque $\vec{\tau}$ at this stage, it is defined to first express the rate of change of angular momentum $\vec{L}$, which is:

$$\vec{\tau} = \left(\frac{d}{dt}\right)\vec{L} = \vec{r}_c \times \vec{F} \tag{3.19}$$

The angular momentum in terms of liner momentum $\vec{p}$ of the particle in motion is expressed as:

$$\vec{L} = \vec{r}_c \times \vec{p} \tag{3.20}$$

For more details of derivation, refer to Chap. 1 under same subject.

Note that, as well, the direction of the magnetic field generated by the gyration is opposite to that of the external field. Thus, the plasma particles that freely are mobile will respond to an external magnetic field with some tendency to reduce the total magnetic field. In other words, plasma is a *diamagnetic* medium and has a tendency to exclude magnetic fields.

As a summary of single-particle motions, so far we covered by applying the general form of Lorenz force Equation 3.2 in uniform electric field $\vec{E}$ and magnetic field $\vec{B}$ by reducing to the form of Equation 3.3 and managed to find the result for a simple harmonic oscillator and consequently the cyclotron frequency as well. In addition, we also found out what the Larmor radius as Equation 3.15 and finally the trajectory of particle function as sets of Equations 3.16a, 3.16b, and 3.16c and showed the concept of the guiding center.

Now we are going to be in quest of all possible forms of general Lorenz force function that will reduce to different categories based on conditions of electric and magnetic field as combined elements of the Lorenz formula.

The sets of Equations 3.16a, 3.16b, and 3.16c also can be written in the following format as a complete set:

$$\begin{aligned} m\dot{v}_x &= qBv_y \quad m\dot{v}_y = -qBv_x \quad m\dot{v}_z = 0 \\ \ddot{v}_x &= \frac{qB}{m}\dot{v}_y = -\left(\frac{qB}{m}\right)^2 v_x \\ \ddot{v}_y &= \frac{qB}{m}\dot{v}_x = -\left(\frac{qB}{m}\right)^2 v_y \end{aligned} \tag{3.21a}$$

The circular orbit around the guiding center (x_0, y_0) which is a fixed point can be written as [3]:

$$\begin{aligned} x - x_0 &= r_L \sin \omega_c t \\ y - y_0 &= \pm r_L \cos \omega_c t \end{aligned} \tag{3.21b}$$

CASE II: Finite $\vec{E}$

In this case, we allow an electric field to be present and the motion to be found as a summation of the two motions, and the usual circular Larmor gyration plus a drift of the guiding center to take place. In this scenario, we take the electric field $\vec{E}$ to lay in the $x - z$ plane; thus, $E_x = 0$. However, the z component of velocity is unrelated to the transverse components as in CASE I above and can be treated separately. Then, the general Lorenz force equation function of motion applies as:

$$\vec{F} = q(\vec{E} + \vec{v} \times \vec{B}) \tag{3.22a}$$

and

$$m\frac{d\vec{v}}{dt} = q(\vec{E} + \vec{v} \times \vec{B}) \tag{3.22b}$$

which has the z component velocity as:

$$\frac{dv_z}{dt} = \frac{q}{m}E_z \tag{3.23a}$$

Integration of Equation 3.23a in respect to time t provides

$$v_z = \frac{qE_z}{m}t + v_0 \tag{3.23b}$$

The above relationships reveal straightforward acceleration along magnetic field $\vec{B}$, and the transverse components of Equations 3.22a and 3.22b will be as:

$$\begin{aligned}
\frac{dv_x}{dt} &= \frac{q}{m}E_x \pm \omega_c v_y \\
\frac{dv_y}{dt} &= 0 \mp \omega_c v_x
\end{aligned} \tag{3.24}$$

Differentiating, we have for constant $\vec{E}$:

$$\begin{aligned}
\ddot{v}_x &= -\omega_c^2 v_x \\
\ddot{v}_y &= \mp \omega_c \frac{q}{m}(E_x \pm \omega_c v_y) = -\omega_c^2 \left(\frac{E_x}{B} + v_y\right)
\end{aligned} \tag{3.25}$$

We can then write the following for this case:

$$\frac{d^2}{dt^2}\left(v_y + \frac{E_x}{B}\right) = -\omega_c^2 \left(v_y + \frac{E_x}{B}\right) \tag{3.26}$$

Comparing this equation with Equations 3.21a and 3.21b, we can easily see that Equation 3.26 is a reduced version of Equations 3.21a and 3.21b as in CASE I if we replace v_y by $v_y + (E_x/B)$. However, the Equations 3.13a and 3.13b therefore can be replaced by:

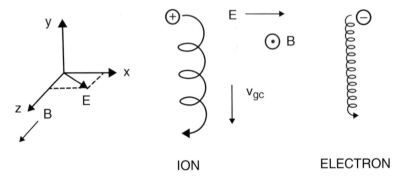

Fig. 3.8 Particle drifts in crossed electric and magnetic fields (Courtesy of Springer Publishing Company) [3]

$$v_x = v_\perp e^{i\omega_c t}$$
$$v_y = \pm i v_\perp e^{i\omega_c t} - \frac{E_x}{B} \tag{3.27}$$

We can find the general form of the Larmor motion as before with the help of superimposition of a drift guiding center velocity $\vec{v}_{gc}$ in the $-y$ direction for $E_x > 0$, which is illustrated in Fig. 3.8 here.

Thus, the general formula by eliminating the term $md\vec{v}/dt$ in Equation 3.22a and doing algebraic homework by taking the vector cross product with the magnetic field, we get:

$$\vec{E} \times \vec{B} = \vec{B} \times (\vec{v} \times \vec{B}) = vB^2 - B(\vec{v} \cdot \vec{B}) \tag{3.28}$$

The transverse components of this equation (i.e., Equation 3.22a) are:

$$\vec{v}_{\perp gc} = \frac{\vec{E} \times \vec{B}}{B^2} \equiv \vec{v}_E \tag{3.29}$$

The magnitude of electric field drift v_E of the guiding center is then given by the following equation as:

$$v_E = \frac{E(V/m)}{B(T)} \frac{m}{s} \tag{3.30}$$

More detailed information and discussion can be found in the book by Chen [3].

CASE III: Nonuniform $\vec{B}$ Field

The above two cases established the concept of the guiding center firmly, and now we need to have some concept and understanding of the particle motion in an inhomogeneous field of electric $\vec{E}$ and magnetic $\vec{B}$ fields where they vary in space or time. Nevertheless, we managed to establish the expression of the guiding center for

uniform fields; however, the problem of the guiding center becomes too complicated to deal with, and we should be able to find exact solution to the problem as
soon as we introduce an inhomogeneity condition to it.

An approximate answer can be found as customary approach to expand in the
small ratio r_L/L, for orbit radius of r_L, where L is the scale length of inhomogeneity.
Seeking for solution using this type of theory called *orbital theory* is extremely
complex and involved, but for the sake of argument, we can study only the simplest
cases as below, where only one inhomogeneity for either electric field or magnetic
one takes place at a time.

CASE III-1: $\nabla \vec{B} \perp \vec{B}$, Gradient $\vec{B}$ Drift

In this case, the magnetic field lines are often called "lines of force," and they are
not lines of force, but they are straight lines, and their density increases as an
example in y-direction as it is illustrated in Fig. 3.9 here.

The solution of this simple case is expressed by Chen [3], and readers should
refer to this book; however, for the sake of this discussion, we summarize the
related equations here, considering the illustration in Fig. 2.9. The gradient in $|\vec{B}|$
does cause the Larmor radius to be larger at the bottom of the orbit than at the top,
and this leads to a drift in opposite directions for ion and electron particles
perpendicular to both $\vec{B}$ and $\nabla \vec{B}$. Under this situation, the drift velocity is
proportional to r_L/L and to $v_\perp$.

For the purpose of this analysis, we consider the Lorentz force $\vec{F} = q\vec{v} \times \vec{B}$
averaged over a gyration, and clearly, since the particle spends more time moving
up and down; thus, $F_x = 0$, as it is shown in Fig. 3.9. Component of Lorentz force in
y-direction, namely, can be calculated F_y in approximation method using the
undisturbed orbit of the particle using Equations 3.13a, 3.13b, and 3.13c to find
the average for a uniform magnetic field $\vec{B}$.

Real part of the complex form of Equations 3.13a and 3.13b is given as:

$$F_y = -qv_x B_z(y) = -qv_\perp(\cos \omega_c t)\left[B_0 \pm r_L(\cos \omega_c t)\frac{\partial B}{\partial y}\right] \tag{3.31}$$

Using Equation 3.31 along with utilization of Taylor series approximation of $\vec{B}$ field
about the point $x_0 = 0$ and $y_0 = 0$, we have:

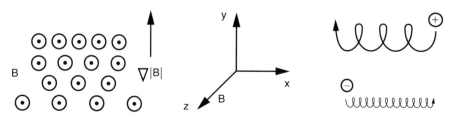

Fig. 3.9 The drift of a gyrating particle in a nonuniform magnetic field (Courtesy of Springer
Publishing Company) [3]

$$\vec{B} = B_0 + (\vec{r} \cdot \nabla)\vec{B} + \cdots$$
$$B_z = B_0 + y(\partial B_z/\partial y) + \cdots \tag{3.32}$$

For this expansion, the required condition $(r_L/L) \ll 1$ needs to hold, where L is the scale length of $\partial B_z/\partial y$. The first term of Equation 3.31 averages to zero in a gyration, and the average of $\cos^2\omega_c t = 1/2$ and then we have:

$$F_y = \mp q v_\perp r_L (\partial B_z/\partial y)/2 \tag{3.33}$$

The guiding center drift velocity then is:

$$\vec{v}_{gc} = \frac{1}{q}\frac{\vec{F} \times \vec{B}}{B^2} = \frac{1}{q}\frac{F_y}{|\vec{B}|}\hat{x} = \mp\left(\frac{v_\perp r_L}{B}\right)\left(\frac{1}{2}\frac{\partial B}{\partial y}\hat{x}\right) \tag{3.34}$$

We have used the following equation in presences of gravitational force by replacing $q\vec{E}$ in the equation motion of Equation 3.22a by the forgoing result that can be applied to the other forces:

$$\vec{v}_f = \frac{1}{q}\frac{\vec{F} \times \vec{B}}{B^2} \tag{3.35}$$

Therefore, we can write the following general form as:

$$\vec{v}_{\nabla B} = \pm\frac{1}{2}v_\perp r_L \frac{\vec{B} \times \nabla\vec{B}}{B^2} \tag{3.36}$$

This equation has all the dependencies that was expected from the physical picture minus the factor 1/2, which is arising from the average.

CASE III-2: Curved $\vec{B}$, Curvature Drift

In this case, we assume the lines of force are curved with a constant radius of curvature R_c and are constant (see Fig. 3.10), and the average square of the component of random velocity $v_{||}^2$ along with centrifugal force F_{cf} is given as:

$$F_{cf} = \frac{mv_{||}}{R_c} = mv_{||}^2\frac{R_c}{R_c^2} \tag{3.37}$$

According to Equation 3.35, this gives rise to a drift:

Fig. 3.10 A curved magnetic field (Courtesy of Springer Publishing Company) [3]

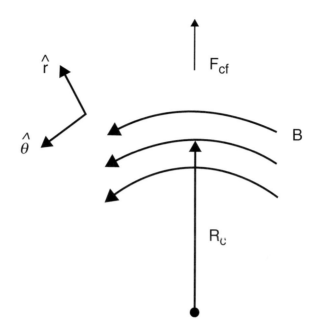

$$\vec{v}_R = \frac{1}{q}\frac{F_{cf} \times \vec{B}}{B^2} = \frac{m v_{\parallel}^2}{qB^2}\frac{\vec{R}_c \times \vec{B}}{\frac{\vec{R}_c}{R_c^2}} \tag{3.38}$$

The general form of total drift in a curved vacuum field:

$$\vec{v}_R + \vec{v}_{\nabla \vec{B}} = \frac{m}{q}\frac{\vec{R}_c \times \vec{B}}{R_c^2 B^2}\left(v_{\parallel}^2 + \frac{1}{2}v_{\perp}^2\right) \tag{3.39}$$

By adding these drifts, which means that if one bends a magnetic field into a torus for the purpose of confining a thermonuclear plasma, the particles will drift out of the torus no matter how one juggles the temperatures and magnetic fields.

For more details and further analysis, readers should refer to Chen's textbook [3].

CASE III-3: $\nabla \vec{B} \| \vec{B}$, Magnetic Mirrors

Now consider magnetic field $\vec{B}$ is primarily laying in the z-direction whose magnetic field varies in that direction and is axisymmetric with $B_{\theta} = 0$ and $\partial/\partial\theta = 0$. Figure 3.11 shows drift of a particle in a magnetic mirror field, where the lines of force cover and diverge with a component of magnetic field B_r in direction r of a cylindrical coordinate. This scenario will give rise to a force, which is trapping a particle in a magnetic field.

We are able to obtain the B_r and $\nabla \cdot \vec{B} = 0$ by the following calculations, using the cylindrical coordinate systems with assumption of axisymmetric around angle θ:

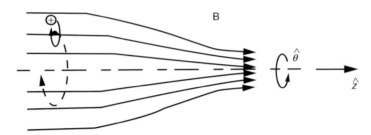

Fig. 3.11 Drift of a particle in a magnetic mirror field (Courtesy of Springer Publishing Company) [3]

$$\frac{1}{r}\frac{\partial}{\partial r}(rB_r) + \frac{\partial B_z}{\partial z} = 0 \tag{3.40}$$

If $\partial B_z/\partial z$ is given at $r = 0$ and does not vary much with r, we have approximately the following:

$$rB_r = -\int_0^r r\frac{\partial B_z}{\partial z}\,dr \simeq -\frac{1}{2}r^2\left[\frac{\partial B_z}{\partial z}\right]_{r=0}$$

$$B_r = -\frac{1}{2}r\left[\frac{\partial B_z}{\partial z}\right]_{r=0} \tag{3.41}$$

The variation of $|\vec{B}|$ with r causes a gradient $\vec{B}$ drift of guiding centers about the axis of symmetry with no radial gradient of magnetic field $\vec{B}$ drift due to $\partial B_\theta/\partial\theta = 0$. Therefore, the components of the Lorentz force are:

$$\begin{aligned}
F_r &= q(v_\theta B_z - v_z B_\theta)\\
F_\theta &= q(-v_r B_z + v_z B_r)\\
F_z &= q(v_r B_\theta - v_\theta B_r)
\end{aligned} \tag{3.42}$$

Moreover, we are interested in the following term of Equation 3.42 as follows:

$$F_z = \frac{1}{2}qv_\theta r\left(\frac{\partial B_z}{\partial z}\right) \tag{3.43}$$

Averaging out this equation over one gyration by considering a particle whose guiding center lies on the axis, then v_θ is a constant during a gyration; depending on the sign of particle charge q, v_θ is $\mp v_\perp$. Since $r = r_{\mathrm{L}}$, the average force is then:

$$F_z = \mp\frac{1}{2}qv_\perp r_{\mathrm{L}}\frac{\partial B_z}{\partial z} = \mp\frac{1}{2}q\frac{v_\perp^2}{\omega_c}\frac{\partial B_z}{\partial z} = -\frac{1}{2}\frac{mv_\perp^2}{B}\frac{\partial B_z}{\partial z} \tag{3.44}$$

Defining the *magnetic moment* of the gyrating particle, which is the same as the definition for the magnetic moment of a current loop with area A and current I showing it as $\mu = IA$, thus we have:

$$\mu = \frac{1}{2}\frac{mv_\perp^2}{B} \qquad (3.45)$$

so that

$$F_z = -\mu\left(\frac{\partial B_z}{\partial z}\right) \qquad (3.46)$$

Then, the general form of force on a diamagnetic particle is as follows:

$$F_{||} = -\mu\left(\frac{\partial B}{\partial s}\right) = -\mu\nabla_{||}B \qquad (3.47)$$

where ds is a line element along the magnetic field $\vec{B}$.

In any case, from the definition for Equation 3.45 and single-particle charge such as ion, e is generated by a charge e coming around $\omega_c/2\pi$ times a second as $I = e\omega_c/2\pi$, and the area A is calculated based on $\pi r_L^2 = \pi v_\perp^2/\omega_c^2$; thus, we can write:

$$\mu = \frac{\pi v_\perp^2}{\omega_c^2}\frac{e\omega_c}{2\pi} = \frac{1}{2}\frac{ev_\perp^2}{\omega_c} = \frac{1}{2}\frac{mv_\perp^2}{B} \qquad (3.48)$$

The Larmor radius varies, as the particle goes through regions of stronger or weaker magnetic field $\vec{B}$; however, the magnetic moment μ does remain *invariant*, and the proof can be seen in Chen's textbook [3].

The invariance of magnetic moment μ is the foundation for one of the initial schemes for plasma confinement approach by the magnetic device called *magnetic mirror*.

Figure 3.12 here shows a simplistic and artistic illustration of such device, where the nonuniform field of a simple pair of coils forms two magnetic mirrors between where the plasma can be trapped as consequently to be confined. This effect works on both ions and electrons, holding either positive or negative charge, respectively.

Conservation of energy requires that:

$$\frac{1}{2}\frac{mv_{\perp 0}^2}{B_0} = \frac{1}{2}\frac{mv_\perp'^2}{B'} \qquad (3.49)$$

where

Fig. 3.12 A plasma trapped
between magnetic mirrors
(Courtesy of Springer
Publishing Company) [3]

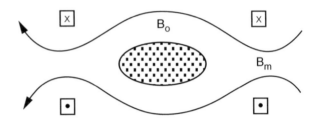

$$v'_\perp 2 = v^2_{\perp 0} + v^2_{\parallel 0} \tag{3.50}$$

Combining Equation 3.49 with Equation 3.50, we can write:

$$\frac{B_0}{B'} = \frac{v^2_{\perp 0}}{v'_\perp 2} = \frac{v^2_{\perp 0}}{v^2_0} \equiv \sin^2\theta \tag{3.51}$$

where θ is the pitch angle of the orbit in the weak field region and with smaller value of this angle; the particle will mirror in regions of higher magnetic field B; however, if this angle is too small, B' exceeds B_m, and the particle does not mirror at all. If we replace B' by B_m in Equation 3.51, we observe that the smallest pitch angle θ of a confined particle is provided by:

$$\sin^2\theta_m = \frac{B_0}{B_m} \equiv \frac{1}{R_m} \tag{3.52}$$

where R_m is the *mirror ratio*.

Figure 3.13 is an illustration of a somewhat called *loss cone*, where Equation 3.52 defines the boundary of a region in velocity space in the shape of cone.

The magnetic mirror first was configured and proposed by Enrico Fermi as an instrument/machine for the acceleration of cosmic rays. His configuration is depicted in Fig. 3.14 here, where protons are bouncing between magnetic fields.

As we stated previously, a further example of the mirror effect confinement of particles can be observed in the Van Allen belts as it was shown in Fig. 3.3.

CASE IV: Nonuniform $\vec{E}$ Field

Now, we assume that, in this case, the magnetic field is uniform and the electric field is in nonuniform conditions, and for simplicity of the problem in hand, we assume electric field $\vec{E}$ is in x-direction and varies in that direction sinusoidally as it is shown in Fig. 3.15 and presented with the following equation:

$$\vec{E} = E_0(\cos kx)\hat{x} \tag{3.53}$$

The associated field distribution has a wavelength $\lambda = 2\pi/k$ and is the result of a sinusoidal distribution of charges, which we do not specify. Practically, such

Fig. 3.13 The loss cone
(Courtesy of Springer
Publishing Company) [3]

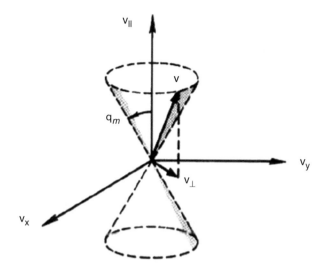

Fig. 3.14 Cosmic ray
proton trap device
(Courtesy of Springer
Publishing Company) [3]

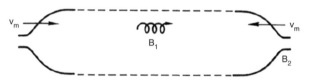

Fig. 3.15 Drift of a
gyrating particle in a
nonuniform electric field
(Courtesy of Springer
Publishing Company) [3]

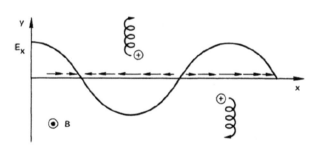

distribution can arise in plasma during a wave motion. Therefore, the equation of
motion is:

$$m\left(\frac{d\vec{v}}{dt}\right) = q\left[\vec{E}(x) + \vec{v} \times \vec{B}\right] \tag{3.54}$$

whose transverse components are:

$$\dot{v}_x = \frac{qB}{m} v_y + \frac{q}{m} E_x(x)$$
$$\dot{v}_y = -\frac{qB}{m} v_x \tag{3.55}$$

and

$$\ddot{v}_x = -\omega_c^2 v_x \pm \omega_c \frac{\dot{E}_x(x)}{B} \tag{3.56}$$

$$\ddot{v}_y = -\omega_c^2 v_y - \omega_c^2 \frac{\dot{E}_x(x)}{B} \tag{3.57}$$

The component of electric field $E_x(x)$ in x-direction in the above equations is presentation of the field at the position of the particle and can be evaluated if we have the knowledge of the particle's orbit, which we need to solve in the first place. However, for a weak electric field, we use an approximation of *undisturbed orbit* to assess $E_x(x)$. The orbit in the absence of the electric field that is given by Equation 3.21b is written as:

$$x = x_0 + r_L \sin \omega_c t \tag{3.58}$$

From Equations 3.57 and 3.53, we obtain:

$$\ddot{v}_y = -\omega_c^2 v_y - \omega_c^2 \frac{\dot{E}_x(x)}{B} \cos k(x_0 + r_L \sin \omega_c t) \tag{3.59}$$

Solution of Equation 3.59 can be found as follows [3]:

$$\ddot{v}_y = 0 = -\omega_c^2 \bar{v}_y - \omega_c^2 \frac{E_0}{B} \overline{\cos k(x_0 + r_L \sin \omega_c t)} \tag{3.60}$$

Expanding the cosine, we have:

$$\cos k(x_0 + r_L \sin \omega_c t) = \cos (kx_0) \cos (kr_L \sin \omega_c t) \\ - \sin (kx_0) \sin (kr_L \sin \omega_c t) \tag{3.61}$$

It will suffice to treat the small Larmor radius case $kr_L \ll 1$. The Taylor expansions are:

$$\cos \varepsilon = 1 - \frac{1}{2} \varepsilon^2 + \cdots$$
$$\sin \varepsilon = \varepsilon + \cdots \tag{3.62}$$

which allows us to write:

$$\cos k(x_0 + r_L \sin \omega_c t) \approx (\cos kx_0)\left(1 - \frac{1}{2}k^2 r_L^2 \sin^2 \omega_c t\right)$$
$$-(\sin kx_0)kr_L \sin \omega_c t$$

(3.63)

Last term of Equation 3.63 vanishes upon averaging over time, and then Equation 3.60 reduces to the following form:

$$\bar{v}_y = -\frac{E_0}{B}(\cos kx_0)\left(1 - \frac{1}{4}k^2 r_L^2\right) = -\frac{E_x(x_0)}{B}\left(1 - \frac{1}{4}k^2 r_L^2\right)$$

(3.64)

Thus, the usual $\vec{E} \times \vec{B}$ drift is modified by the inhomogeneity to read:

$$\vec{v}_E = \frac{\vec{E} \times \vec{B}}{B^2}\left(1 - \frac{1}{4}k^2 r_L^2\right)$$

(3.65)

Chen [3] argues about finding the *finite-Larmor-radius effect*, using the expansion of Equation 3.65 as a following form, and readers should refer to that reference:

$$\vec{v}_E = \left(1 - \frac{1}{4}r_L^2 \nabla^2\right)\frac{\vec{E} \times \vec{B}}{B^2}$$

(3.66)

CASE V: Time-Varying $\vec{E}$ Field

In this case, we just use the equation related to the case and leave all the details to the reader to see the proof of the details in Chen [3] and other plasma-related books. The condition that we consider this case calls for both electric and magnetic to be uniform in space but varying in time.

$$\vec{E} = E_0 e^{i\omega t}\hat{x}$$

(3.67)

Since $\dot{E}_x = i\omega E_x$, we can write Equation 3.56 as:

$$\ddot{v}_x = -\omega_c^2\left(v_x \mp \frac{i\omega}{\omega_c}\frac{\tilde{E}_x}{B}\right)$$

(3.68)

Let us write the rest of the equation related to this case just as they are without any detailed explanations:

$$\tilde{v}_p \equiv \pm \frac{i\omega \tilde{E}_x}{\omega_c B}$$

$$\tilde{v}_E \equiv \frac{\tilde{E}_x}{B}$$

(3.69)

$$\ddot{v}_x = -\omega_c^2 \left(v_x - \tilde{v}_p \right)$$
$$\ddot{v}_y = -\omega_c^2 \left(v_y - \tilde{v}_E \right)$$

(3.70)

Solution of Equation 3.70 is:

$$v_x = v_\perp e^{i\omega_c t} + \tilde{v}_p$$
$$v_y = \pm i v_\perp e^{i\omega_c t} + \tilde{v}_E$$

(3.71)

Twice differentiation of Equation 3.71 in respect to time results in:

$$\ddot{v}_x = -\omega_c^2 v_x + \left(\omega_c^2 - \omega^2 \right) \tilde{v}_p$$
$$\ddot{v}_y = -\omega_c^2 v_y + \left(\omega_c^2 - \omega^2 \right) \tilde{v}_E$$

(3.72)

Polarization drift for x-component along the direction of $\vec{E}$ field is given as:

$$\vec{v}_p = \pm \frac{1}{\omega_c \vec{B}} \frac{d\vec{E}}{dt}$$

(3.73)

In addition, polarization current is:

$$\vec{j}_p = ne \left(v_{ip} - v_p \right) = \frac{ne}{eB^2} (M + m) \frac{d\vec{E}}{dt} = \frac{\rho}{B^2} \frac{d\vec{E}}{dt}$$

(3.74)

where ρ is the mass density, while M and m are particle masses involved, and they are defined as before.

If a field $\vec{E}$ is suddenly applied, the first thing the ion does is to move in the direction of $\vec{E}$. Only after picking up a velocity $\vec{v}$ does the ion feel a Lorentz force $e\vec{v} \times \vec{B}$ and begin to move downward, as it is illustrated in Fig. 3.16.

CASE VI: Time-Varying $\vec{B}$ Field

For this case, we let the magnetic field vary in time and due to the fact that the Lorentz force is perpendicular to $\vec{v}$, a magnetic field by itself does not have any impact energy to a charged particle. However, associated with magnetic field $\vec{B}$, there exists an electric field $\vec{E}$ that is given as below that can accelerate the particle:

$$\frac{\nabla \times \vec{E} = -\dot{\vec{B}}}{B^2}$$

(3.75)

Fig. 3.16 The polarization
drift (Courtesy of Springer
Publishing Company) [3]

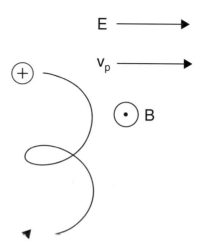

Details this analysis has worked out by Ref. [3] and we briefly show all the related
equations, including the magnetic moment μ that is invariant or is slowly varying
magnetic fields and magnetic flux Φ through a Larmor orbit that is constant as:

$$
\delta\left(\frac{1}{2}mv_\perp^2\right) = \mu\delta B
$$

$$
\delta\mu = 0 \qquad = B\pi\frac{v_\perp^2}{q^2B^2} = \frac{2\pi m\frac{1}{2}mv_\perp^2}{q^2}\frac{1}{B} = \frac{2\pi m}{q^2}\mu \tag{3.76}
$$

$$
\Phi = B\pi\frac{v_\perp^2}{\omega_c^2}
$$

This property is used in a method of plasma heating known as adiabatic compression. Figure 3.17 shows a schematic of how this is done. A plasma is injected into
the region between the mirrors A and B. Coils A and B are then pulsed to increase B
and hence $v_\perp^2$. The heated plasma can then be transferred to the region C-D by a
further pulse in A, increasing the mirror ratio there. The coils C and D are then
pulsed to further compress and heat the plasma. Early magnetic mirror fusion
devices employed this type of heating [3].

3.2.1   Summary of the Guiding Center Drift

$$
\text{General force } \vec{F}: \quad \vec{v}_f = \frac{1}{q}\frac{\vec{F}\times\vec{B}}{B^2} \tag{3.77}
$$

$$
\text{Electric field } \vec{E}: \quad \vec{v}_E = \frac{\vec{E}\times\vec{B}}{B^2} \tag{3.78}
$$

Fig. 3.17 Two-stage
adiabatic compression of
plasma (Courtesy of
Springer Publishing
Company) [3]

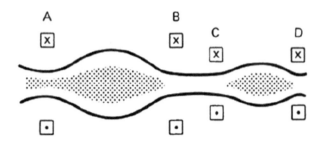

$$\text{Gravitational field } \vec{v}_g : \quad \vec{v}_g = \frac{m}{q} \frac{\vec{g} \times \vec{B}}{B^2} \tag{3.79}$$

$$\text{Nonuniform } \vec{E} : \quad \vec{v}_E = \left(1 + \frac{1}{4} r_L^2 \nabla^2 \right) \frac{\vec{E} \times \vec{B}}{B^2} \tag{3.80}$$

Nonuniform magnetic field $\vec{B}$

$$\text{Grad-}\vec{B}\text{ drift}: \quad \vec{v}_{\nabla \vec{B}} = \pm \frac{1}{2} v_\perp r_L \frac{\vec{B} \times \nabla \vec{B}}{B^2} \tag{3.81}$$

$$\text{Curvature drift}: \quad \vec{v}_R = \frac{m v_\parallel^2}{q} \frac{\vec{R}_c \times \vec{B}}{R_c^2 B^2} \tag{3.82}$$

$$\text{Curved vacuum field}: \quad \vec{v}_R + \vec{v}_{\nabla \vec{B}} = \frac{m}{q} \left(v_\parallel^2 + \frac{1}{2} v_\perp^2 \right) \frac{\vec{R}_c \times \vec{B}}{R_c^2 B^2} \tag{3.83}$$

$$\text{Polarization drift}: \quad \vec{v}_p = \pm \frac{1}{\omega_c \vec{B}} \frac{d\vec{E}}{dt} \tag{3.84}$$

However, more details can be in a lot of standard plasma textbooks.

3.3   How the Tokamak Reactors Works

The tokamak was invented in the old Soviet Union by Andrei Sakharov and Igor
Tamm, and basically, the artistic configuration of it is shown here in Fig. 3.18.

As of 2008, the US Department of Energy (DOE) and other US federal agencies
have spent approximately 18 billion dollars on energy devices using the fusion
reaction between deuterium and tritium (D-T Fusion, below the left of Fig. 3.19). In
this reaction the hydrogen isotope deuterium (with one "extra" neutron) collides
with the hydrogen isotope tritium (with two "extra" neutrons) to form an alpha
particle (a helium nucleus) and a neutron. This is a nuclear reaction: between them,
the new alpha and the neutron possess 17.6 MeV (million electron volts) of energy.

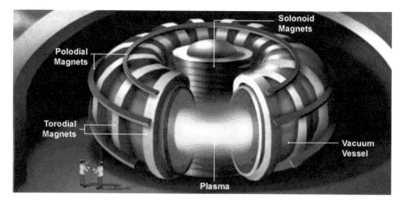

Fig. 3.18 Conceptual sketch of tokamak

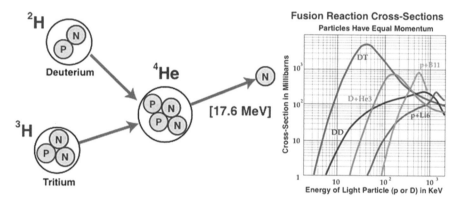

Fig. 3.19 Depiction of all isotopes of hydrogen thermonuclear reactions

In the Fusion Reaction Cross-Sections graph (above the right in Fig. 3.19), the red deuterium-tritium (D-T) curve peaks at about 40 KeV (40,000 eV). This means that the optimum activation energy required for the D-T fusion reaction is only about 40 KeV. The curves for the other reactions peak at much higher energies. The energy required to make the D-T reaction happen is lower (in KeV) than the energy required for any other nuclear fusion reaction. In addition, the height of the D-T curve (cross section in millibarns) indicates that the deuterium and tritium isotopes "see" each other as being relatively large, compared to the isotopes in the other reactions shown. Thus, at the proper activation energy, this reaction is much more likely to happen than any other fusion reaction. DOE and many other entities pursue the D-T reaction because it requires less energy to initiate and because it is more probable.

Unfortunately, there are several serious disadvantages to this reaction:

1. Tritium is both radioactive and expensive.

2. The neutrons released can harm living things and damage any other materials surrounding them.
3. The neutrons can make some materials radioactive.

At this time, the device preferred for making this reaction happen is the tokamak. The DOE, the European Union, Japan, Russia, China, and India are all part of the International Thermonuclear Experimental Reactor (ITER) program, which is working on it. Their dream is that the tokamak will heat a plasma containing tritium and deuterium nuclei. The hotter these nuclei get, the faster they will move. When the plasma is hot enough, some of the nuclei will be moving fast enough to react when they collide. The energy of the newly produced, highly energetic helium nuclei (alphas) will be used to keep the plasma hot, and the energy of the new neutrons will be released to a lithium metal blanket, which lines the tokamak. Water lines will run through the lithium. The hot lithium will heat the water to steam, and the steam will be used to spin turbines, which will spin generators to make electricity.

There is a substantial gap between the above dream and its fulfillment. For at least 50 years, the practical use of tokamaks and other D-T devices to make electricity has been forecast to be "about 30 years in the future." To be commercially useful, a controlled fusion reaction must produce more energy than the energy that was required to cause the reaction in the first place (i.e., the 40 KeV activation energy is mentioned above). The point at which the energy produced exceeds the energy required is called "net power" or "break-even." Various organizations in different parts of the world have been working to produce "net power" nuclear fusion for about 50 years. Many billions of rubles, dollars, yen, and euros have been spent on this endeavor, but no one has been successful yet.

Many of the efforts have involved the idea of heating plasma of deuterium (D) and tritium (T) gases until the nuclei fuse. When the heat of plasma increases, the average energy (speed) of the particles increases, but there is an enormous variation in the energies of the individual particles within the plasma. This set of all the different energies of the particles in plasma or gas is called a Maxwellian distribution. Unfortunately, in the typical Maxwellian distribution, only few of the nuclei have the 40 KeV of energy required to react, and all the other particles are just going along for the ride. If the temperature is increased to the point where an adequate number of nuclei have enough energy, then other problems develop which can compromise the integrity of the containment.

Both the tokamak and the stellarator use magnetic fields to manipulate the D-T plasma. However, the distinguishing feature of the tokamak is its "step-down" transformer. The transformer's primary is the stack of beige coils in the center of the tokamak's torus (in the donut's hole below, Fig. 3.20). The transformer's secondary is the ring of the plasma—the orange skinny donut. An increasing current in the many-coiled primary induces a much larger current in the single-coiled plasma "donut" secondary.

Two magnetic fields combine to produce the resultant magnetic field (labeled left) that spirals helically around the tokamak's torus (orange skinny donut). This

Fig. 3.20 Tokamak donut
hole shape

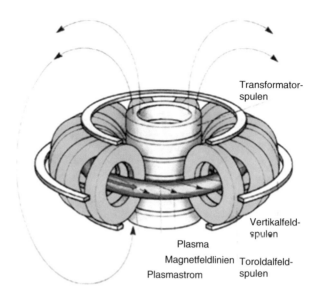

Transformator-
spulen

Vertikalfeld-
spulen

Plasma

Magnetfeldlinien Toroldalfeld-

Plasmastrom   spulen

resultant field contains and controls the plasma. The two magnetic fields that combine vectorially to make the resultant field are (1) the toroidal field, generated by the green **toroidal** coils, and (2) the **poloidal** field generated by the orange plasma current in the torus. The vertical coils (the large rings around the outside of the tokamak and above and below it) can create a vertical magnetic field for controlling the position of the plasma inside the torus.

The transformer coils also cause "ohmic" (RI^2) heating in the plasma, which contributes to raising its temperature. However, since the electrical resistance of the plasma decreases as its temperature increases, the upper limit on the "ohmic" heating turns out to be about 20–30 million degrees Celsius, which is not high enough for fusion.

Thus it is necessary to further increase the temperature by three additional strategies: radio-frequency heating, magnetic compression, and neutral beam injection.

The proposed ITER tokamak, to be built in France, is pictured in Fig. 3.21. To get an idea of the scale that is involved, notice the tiny little lab tech in the blue coat standing on the floor, near the machine.

A somewhat similar fusion effort is the stellarator, also known as the Wendelstein 7-X in Germany (see Fig. 3.22).

Both the stellarator and the tokamak use a magnetic containment to control the fuel. A distinguishing feature of the stellarator is the use of odd-shaped coils to manipulate the shape of the plasma donut within the coils. To have a better concept of how the stellarator works, we introduce the plasma beta β, which is the ratio of plasma pressure to magnetic pressure and is defined as:

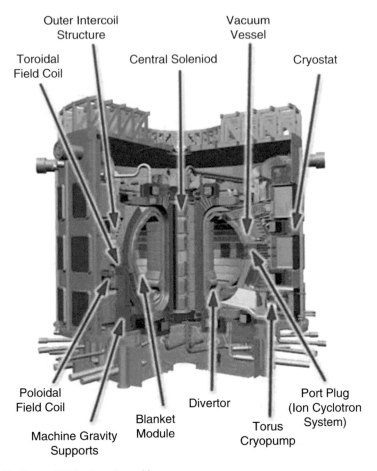

Outer Intercoil Structure

Toroidal Field Coil

Central Soleniod

Vacuum Vessel

Cryostat

Poloidal Field Coil

Divertor

Port Plug (Ion Cyclotron System)

Blanket Module

Machine Gravity Supports

Torus Cryopump

Fig. 3.21 France ITER tokamak machine

$$\beta = \frac{P_{\text{Plasma}}}{p_{\text{Magnetic}}} = \frac{nk_{\text{B}}T}{B^2/(2\mu_0)} \tag{3.85}$$

where

$n = $ Plasma Density

$k_{\text{B}} = $ Boltzmann Constant

$T = $ Plasma Temperature

$B = $ Magnetic Field

$\mu_0 = $ Magnetic Moment

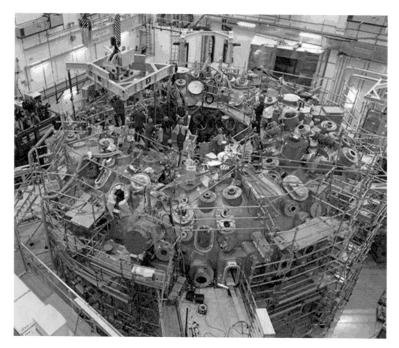

Fig. 3.22 Stellarator, Wendelstein 7-X under construction in Germany

Given that the magnets are a dominant factor in magnetic fusion confinement (MFC) reactor design and that density and temperature combine to produce pressure, the ratio of the pressure of the plasma to the magnetic energy density naturally becomes a useful figure of merit when comparing MCF designs. In effect, the ratio illustrates how effectively a design confines its plasma.

β is normally measured in terms of the total magnetic field and the term is commonly used in studies of the Sun and Earth's magnetic field and in the field of magnetic fusion power designs. However, in any real-world design, the strength of the field varies over the volume of the plasma, so to be specific; the average beta is sometimes referred to as the "beta toroidal." In the tokamak design, the total field is a combination of the external toroidal field and the current-induced poloidal one, so the "beta poloidal" is sometimes used to compare the relative strengths of these fields. In addition, as the external magnetic field is the driver of reactor cost, "beta external" is used to consider just this contribution.

In the magnetic fusion power field, plasma is often confined using large superconducting magnets that are very expensive. Since the temperature of the fuel scales with pressure, reactors attempt to reach the highest pressures possible. The costs of large magnets roughly scale like $\beta^{1/2}$. Therefore beta can be thought of as a ratio of money out to money in for a reactor, and beta can be thought of (very approximately) as an economic indicator of reactor efficiency. To make an economically useful reactor, betas better than 5 % are needed.

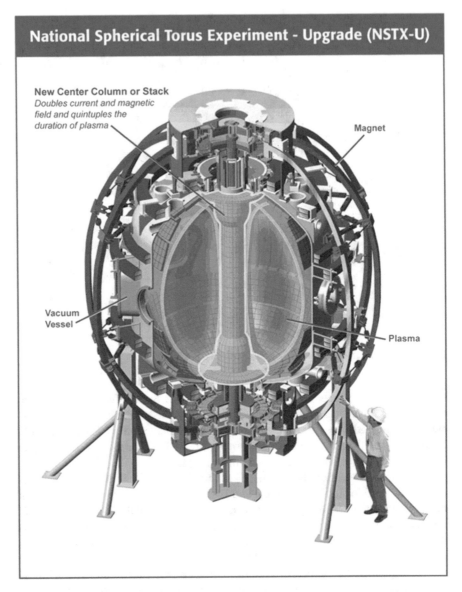

Fig. 3.23 Illustration of spherical tokamak

The same term is also used when discussing the interactions of the solar wind with various magnetic fields. For example, beta in the corona of the Sun is about 0.01.

Back to our original discussion, tokamaks have been studied the most and have achieved the best overall performance for MCF purposes; however, the stellarator is followed by the spherical tokamak (see Fig. 3.23), which is actually a very tight aspect ratio tokamak.

These configurations (i.e., tokamak and stellarator) all have relatively strong toroidal magnetic fields and reasonable transport losses. Each is capable of MHD stable operation at acceptable values of β, without the need for a conducting wall close to the plasma.

The advantage of the stellarator is that only concept, which does not require toroidal current device in a magnetic plasma fusion reactor but has a noticeably more complicated magnetic configuration which increases complexity and cost.

3.4 Intertial Confinement

Inertial confinement fusion (ICF) in recent years has raised a lot of interest beyond just the national laboratories in the USA and abroad. ICF's aim is toward producing clean energy, using high-energy laser beam or for that matter a particle beam (i.e., the particle beam may consist of heavy or light ion beam) to drive a pellet of two isotopes of hydrogen to fuse and release energy. See the D-T Fusion process in Equations 3.23a and 3.23b, where n is the neutron and α is the particle such as helium ($^4_2$He):

$$D + T \rightarrow n(14.06\,\text{MeV}) + \alpha(3.52\,\text{MeV}) \tag{3.86}$$

These two isotopes of hydrogen are known as deuterium (D=^{2}H) and tritium (T=^{3}H) as part of raising the fuel to ignition temperature in order to satisfy the confinement criteria of $\rho r \geq 1$ g/cm^2, where ρ and r are the compressed fuel density and radius pellet, respectively. In order for the confinement criteria, also known as the Lawson criterion, to be satisfied, it needs to take place before the occurrence of the Rayleigh-Taylor hydrodynamics instability would happen for uniform illumination of the target's surface, namely, the pellet of deuterium and tritium.

In a direct laser-driven pellet approach, in order to overcome the Raleigh-Taylor instability, we require a large number of laser beams (see Fig. 3.24).

Fig. 3.24 is the schematic of the stages of inertial confinement fusion using lasers. The blue arrows represent radiation; orange is blowoff; purple is inwardly transported thermal energy.

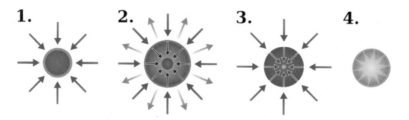

Fig. 3.24 Direct laser-driven compression of a fusion pellet

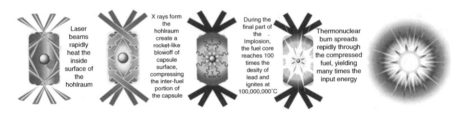

Fig. 3.25 Indirect soft X-ray hohlraum drive compression of fusion pellet

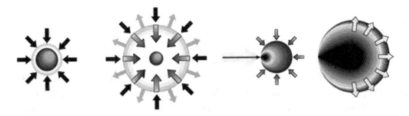

Fig. 3.26 Single-beam igniter concept for fusion pellets

1. Laser beams or laser-produced X-rays rapidly heat the surface of the fusion target, forming a surrounding plasma envelope.
2. Fuel is compressed by the rocket-like blowoff of the hot surface of the material.
3. During the final part of the capsule implosion, the fuel core reaches 20 times the density of lead and ignites at 100,000,000 °C.
4. Thermonuclear burn spreads rapidly through the compressed fuel, yielding many times the input energy.

In case of indirect illuminating target approach, the laser light is converted into soft X-ray, which is trapped inside a hohlraum chamber surrounding the fusion fuel irradiating it uniformly. In this approach, in order to archive fusion inertial confinement, the energy source that drives the ablation and compression as it was stated is soft X-ray ratio. This is produced by the conversion of a nonthermal, directed energy source, such as lasers or ion beams, into thermal radiation inside a high-opacity enclosure that is referred to as a hohlraum (see Fig. 3.25).

Fig. 3.26 is the schematic of the stages of inertial confinement fusion using lasers to drive the pellet, the compression proceeds along several steps from left to right as:

1. Laser illumination: Laser beam rapidly heats the inside surface of the hohlraum.
2. Indirect drive illumination: The walls of the hohlraum create an inverse rocket effect from the blowoff of the fusion pellet surface, compressing the inner fuel portion of the pellet.
3. Fuel pellet compression: During the final part of the implosion process, the fuel core reaches a high density and temperature.
4. Fuel ignition and burn: The thermonuclear burn propagates through the compressed fusion fuel amplifying the input energy in fusion fuel burnup.

In addition to the above approaches, there is a third approach as it is depicted in Fig. 3.26, and that is a single-beam direct approach, where a single beam is used for the compression along the following steps:

1. Atmospheric formation: A laser or a particle beam rapidly heats up the surface of the fusion pellet surrounding it with a plasma envelope.
2. Compression: The fuel is compressed by the inverse rocket blowoff of the pellet surface imploding it inward.
3. Beam fuel ignition: At the instant of maximum compression, a short high-intensity pulse ignites the compressed core. An intensity of 10^{19} [W/cm^2] is contemplated with a pulse duration of 1–10 μs.
4. Burn phase: The thermonuclear burn propagates through the compressed fusion fuel yielding several times the driver input energy.

In either approaches above, the Lawson criterion for a simple case of physics of inertial confinement fusion (ICF) can easily be calculated as follows.

The Lawson criterion applies to inertial confinement fusion (ICF) as well as to magnetic confinement fusion (MCF) but is more usefully expressed in a different form. A good approximation for the inertial confinement time τ_E is the time that it takes an ion to travel over a distance r at its thermal speed $v_{Thermal}$.

$$v_{Thermal} = \sqrt{\frac{k_B T}{m_i}} \qquad (3.87)$$

where:

$$k_B = \text{is Boltzmann constant}$$
$$m_i = \text{is Mean ionic mass}$$
$$T = \text{is Temperature}$$

Equation 3.87 is derived from kinetic energy theory and gas pressure relationship. The inertial confinement time τ_E can thus be approximated as:

$$\tau_E \approx \frac{r}{v_{Thermal}} \qquad (3.88)$$

Substituting Equation 3.87 into Equation 3.88 results in:

$$\tau_E \approx \frac{r}{v_{\text{Thermal}}}$$

$$= \frac{r}{\sqrt{\dfrac{k_B T}{m_i}}} \tag{3.89}$$

$$= r \cdot \sqrt{\frac{m_i}{k_B T}}$$

However, the Lawson criterion requires that fusion heating fE_{ch} exceeds the power losses P_{loss} as written below:

$$fE_{ch} \geq P_{loss} \tag{3.90}$$

In this equation, the *volume rate* is f, which is the reactions per volume time of fusion reaction and is written as:

$$f = n_{\text{Deuterium}} n_{\text{Tritium}} <\sigma v> = \frac{1}{4} n^2 <\sigma v> \tag{3.91}$$

Moreover, E_{ch} is the energy of the charged fusion products, and in case of deuterium-tritium, reaction is equal to 3.5 MeV.

In addition, power loss density P_{loss} is the rate of emery loss per unit volume and is written as:

$$P_{loss} = \frac{W}{\tau_E} \tag{3.92}$$

where W is the energy density or energy per unit volume and is given by:

$$W = 3nk_B T \tag{3.93}$$

In all above equations, the variables that are used are defined as below:

k_B = Boltzmann constant

n = Particle density

$n_{\text{Deuterium}}$ = Deuterium particle density

n_{Tritium} = Tritium particle density

τ_E = Confinement time that measure the rate at which a system loses energy to its sounding environment

σ = Fusin cross section

v = Relative velocity

$<\sigma v>$ = Average over the Maxwellian velocity distribution at temperature T

T = Temperature

Now substituting for all the quantities in Equation 3.90, the result is written as:

$$n\tau_E \geq \frac{12}{E_{ch}} \frac{k_B T}{< \sigma v >} \equiv L \qquad (3.94)$$

Equation 3.95 is known as Lawson criterion and for the deuterium and tritium reaction is at least $n\tau_E \geq 1.5 \times 10^{20} s/m^3$, where the minimum of the product occurs near $T = 25$ keV. The quantity $T/< \sigma v >$ is a function of temperature with an absolute minimum. Replacing the function with its minimum value provides an absolute lower limit for the product $n\tau_E$.

Substituting Equation 3.95 into Equation 3.89, we obtain:

$$n\tau_E \approx n \cdot r \cdot \sqrt{\frac{m_i}{k_B T}} \geq \frac{12}{E_{ch}} \frac{k_B T}{< \sigma v >} \qquad (3.95)$$

or

$$n \cdot r \geq \frac{12}{E_{ch}} \cdot \frac{(k_B T)^{3/2}}{< \sigma v > \cdot m_i^{1/2}} \qquad (3.96)$$

Equation 3.97 could be, approximated to the following form as:

$$n \cdot r \geq \frac{(k_B T)^{3/2}}{< \sigma v >} \qquad (3.97)$$

This product must be greater than a value to the minimum of $T^{3/2}/< \sigma v >$. The same requirement is traditionally expressed in terms of mass density $\rho =< n m_i >$ as:

$$\rho r \geq 1 g/cm^2 \qquad (3.98)$$

Satisfaction of this criterion at the density of solid deuterium-tritium (0.2 g/cm^3) would require a laser pulse of implausibly large energy. Assuming the energy required scales with the mass of the fusion plasma ($E_{laser} \approx \rho r^3 \approx \rho^{-2}$), compressing the fuel to 10^3 or 10^4 times solid density would reduce the energy required by a factor of 10^6 or 10^8, bringing it into a realistic range. With a compression by 10^3, the compressed density will be 200 g/cm^3, and the compressed radius can be as small as 0.05 mm. The radius of the fuel before compression would be 0.5 mm. The initial pellet will be perhaps twice as large, since most of the mass will be ablated during the compression.

The fusion power density is a good figure of merit to determine the optimum temperature for magnetic confinement, but for inertial confinement, the fractional burnup of the fuel is probably more useful. The burnup should be proportional to the specific reaction rate ($n^2 < \sigma v >$) times the confinement time (which scales as $T^{-1/2}$) divided by the particle density n:

$$\text{burn-up fraction} \quad \Rightarrow \quad \begin{cases} \propto & n^2 < \sigma v > T^{-1/2}/n \\ \propto & (nT) < \sigma v > /T^{3/2} \end{cases} \tag{3.99}$$

Thus the optimum temperature for inertial confinement fusion maximizes $< \sigma v > /T^{3/2}$, which is slightly higher than the optimum temperature for magnetic confinement.

Note that as part of key issues, for the laser to drive the pellet of micro-balloon containing deuterium and tritium to achieve fusion is a symmetrical homogenous compression, which means targeting for a perfectly spherical implosions and explosions. However, in reality, this ideal situation never will take place to its perfection and as a result has a number of physics, problem consequences and they are:

- Instabilities and Mixing

  - Rayleigh-Taylor unstable compression
  - Break of symmetry destroys confinement.

- How to improve energy coupling into target (i.e., pellet of D-T), which requires the conversion of kinetic energy from the implosion into internal energy of the fuel that is not perfect. Additionally, we need to prevent the reduction of the maximum compression.
- Severe perturbing of a spherical homogeneous and symmetric implosion can result in small-scale turbulences and even to the breakup of the target shell.
- The hot-spot area at the ablation surface is increased or has a large surface due to the perturbed structure, which leads to reduction of ignition temperature to achieve fusion reaction in the corona of the pellet, and it causes the α-particle created in Equation 3.86 to escape the hot-spot area. This also lowers the self-heating (see Figs. 3.27 and 3.28 below)
- Finally, what is the best material for the first wall of the pellet of D-T as a target?

In summary, the Rayleigh-Taylor instabilities occur when a lower density fluid such as oil underlies a higher density fluid such as water. In inertial confinement where the implosion and explosion process takes place in a sequence, the higher density fluid is the pellet surface, and the lower density fluid is the plasma surrounding it and compressing the pellet through the inverse rocket action (i.e., inertial) of the implosion process.

In all approaches stated above, the inertial confinement via laser or particle beams imploding and exploding the target pellet in a symmetrical and homogeneous mode is mainly influenced by Rayleigh-Taylor (RT) instabilities at the ablation surface.

As we stated, the impact and effect of the Rayleigh-Taylor (RT) instabilities is because they initially grow exponentially, so that even very small and insignificant disturbances can grow to a size that has adverse effect on the entire compression in a homogeneous and symmetrical mode, as it is observed in Fig. 3.29 illustration.

Fig. 3.27 Depiction of the reduced self-heating of the hot-spot area from prematurely escaping α-particles

Fig. 3.28 Striking similarities exist between hydrodynamic instabilities in: (**a**) inertial confinement fusion capsule implosions and (**b**) core-collapse supernova explosions

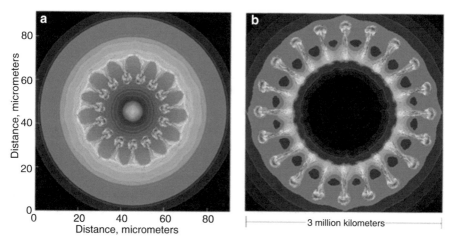

In this illustration, the major instability is again because of heavy material pushed on a low-density one.

This instability always occurs, since the laser or particle beam as the driver of the deuterium-tritium pellet is never 100 % homogeneous and symmetric; consequently, the Rayleigh-Taylor instability always is growing.

The growth rate of the Rayleigh-Taylor instability can be measured in a wavelength range not previously accessible, and it is very important factor that one needs

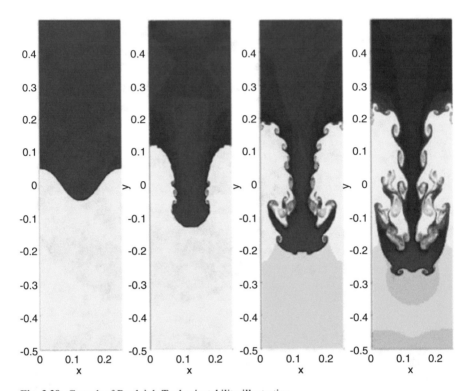

Fig. 3.29 Growth of Rayleigh-Taylor instability illustration

to pay attention to it during the implosion and explosion of the pellet. Moreover, it is important for this purpose to deliver energy to the corona of the pellet as symmetrically and homogenously as possible before the plasma frequency generated at the ablation surface reaches the beam wavelength frequency as the driver (see Fig. 3.30).

Thus, in conclusion, the fusion targets can be illuminated with the energy of different drivers. The primary efforts in inertial confinement exist in the USA, France, and Japan.

The National Ignition Facility (NIF) is the world's largest and most energetic laser facility ever built. NIF is also the most precise and reproducible laser as well as the largest optical instrument. The giant laser has nearly 40,000 optics, which precisely guide, reflect, amplify, and focus 192 laser beams onto a fusion target about the size of a pencil eraser. NIF became operational in March 2009. NIF is the size of a sports stadium—three football fields could fit inside. In Fig. 3.31 is the artistic top view of this facility at Lawrence Livermore National Laboratory (LLNL) in California.

NIF is making important advances toward achieving fusion ignition in the laboratory for the first time. NIF's goal is to focus the intense energy of 192 giant laser beams on a BB-sized target (see Fig. 3.32) filled with hydrogen fuel, fusing the

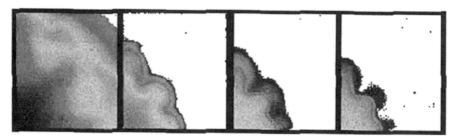

Fig. 3.30 Growth of Rayleigh-Taylor instabilities during pellet implosion

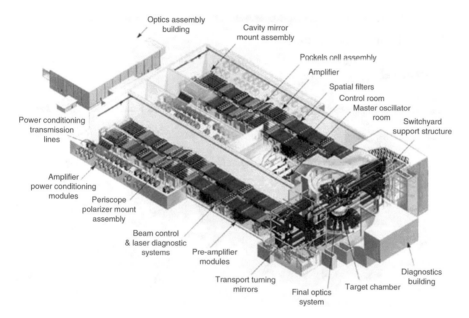

Fig. 3.31 Top view of National Ignition Facility (NIF) at Lawrence Livermore Laboratory

hydrogen atoms' nuclei and releasing many times more energy than it took to initiate the fusion reaction.

In addition to energy for the future, the NIF's primary missions include national security and understanding the universe. Moses noted that there are many benefits to being able to create conditions to study fusion reactions in lieu of weapons testing. The NIF target chamber will also allow the study of the cosmos, for example.

BB-sized target is a hohlraum cylinder, which contains the NIF fusion fuel capsule, which is just a few millimeters wide, about the size of a pencil eraser, with beam entrance holes at either end. The fuel capsule is the size of a small pea.

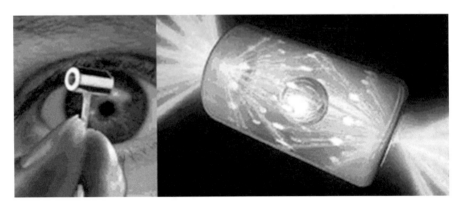

Fig. 3.32 Illustration of a BB-sized target

References

1. J.D. Lawson, Some criteria for a power producing thermonuclear reactor. Proc. Phys. Soc. **70**, 6 (1957)
2. U.S. Inan, M. Golkowski, *Principles of Plasma Physics for Engineers and Scientists* (Cambridge University Press, Cambridge, 2011)
3. F.F. Chen, *Introduction to Plasma Physics and Controlled Fusion*, 3rd edn. (Springer Publishing Company, New York, 2015)

Index

© Springer International Publishing AG 2016
B. Zohuri, *Plasma Physics and Controlled Thermonuclear Reactions Driven Fusion Energy*, DOI 10.1007/978-3-319-47310-9

Printed in the United States
By Bookmasters